CCEA GCSE GEOGRAPHY

THIRD EDITION

Orders: please contact Hachette UK Distribution, Hely Hutchinson Centre, Milton Road, Didcot, Oxfordshire, OX11 7HH. Telephone: +44 (0)1235 827827. Email education@hachette.co.uk Lines are open from 9 a.m. to 5 p.m., Monday to Friday. You can also order through our website: www.hoddereducation.co.uk

ISBN: 9781471891687

First published in 2017 by

Hodder Education,

An Hachette UK Company

Carmelite House

50 Victoria Embankment

London EC4Y 0DZ

www.hoddereducation.co.uk

Impression number 10 9

Year 2023

Cover photo © NASA/Science Photo Library

Illustrations by Barking Dog Art

Typeset by Hart McLeod Ltd.

Printed and bound by CPI Group (UK) Ltd, Croydon, CR0 4YY

A catalogue record for this title is available from the British Library.

CCEA

GCSE GEOGRAPHY

THIRD EDITION

Petula Henderson
Stephen Roulston

CONTENTS

UNIT 1

UNDERSTANDING OUR NATURAL WORLD

THEME A: River Environments

Niagara Falls, on the border between Canada and the USA.

How do rivers manage to create such a dramatic landform?

1 The drainage basin: a component of the water cycle

By the end of this section you will be able to:

- identify and define the characteristics of a drainage basin
- know and understand the elements of a drainage basin
- know and understand inputs, stores and transfers.

Tip

Make sure you know what all the main characteristics of a drainage basin are. Try using flash cards to help you revise them regularly.

Characteristics of a drainage basin

What are the characteristics of a drainage basin?

Water is a critical resource. The water that is most useful to humans is fresh water, although this only makes up 2.5 per cent of all the water on the planet, and only 0.1 per cent is stored in rivers and lakes. The rest of the fresh water on land is stored in ice sheets and glaciers, or in the soil and deeper down in the ground. The total amount of water on our planet never changes: in other words, none arrives from space, and none is lost to space. This is called a closed system. The water on Earth circulates between the sea, land and air (stores), being recycled in a natural process known as the hydrological cycle (water cycle).

On the land, water is stored on the surface as lakes and rivers. Each river is contained within its own drainage basin. This is the area of land drained by a river, from its source (where it begins) to its mouth (where it ends by meeting the sea, ocean or lake) and its tributaries. The boundary of a drainage basin follows a ridge of high ground, known as the watershed. A confluence is when a tributary meets the main river. In some cases, confluences can be very clear if the nature of the water in the two meeting rivers differs, like the confluence in Brazil where the River Negro and the River Solimoes meet. This and other features of the drainage basin are summarised in Figure 2.

Figure 1 The confluence of a river.

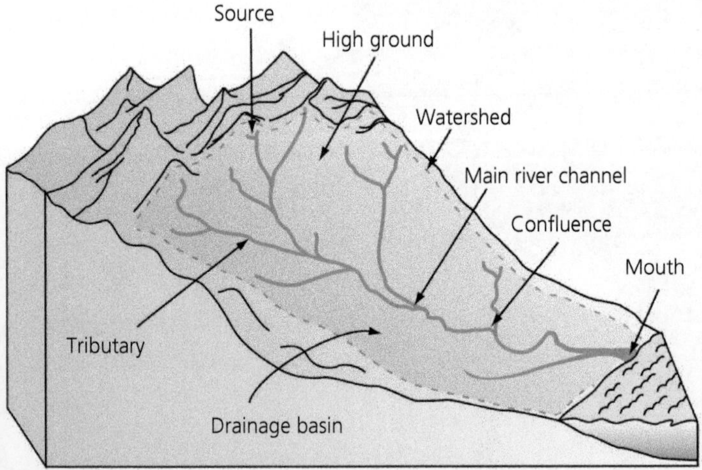

Figure 2 A generalised drainage basin.

The amount of water within a single drainage basin can vary, as it has inputs (from precipitation) and outputs (from evapotranspiration). Therefore this is an open system.

Figure 3 A simple open system of the drainage basin.

What are the components of a drainage basin, and how do they relate to one another?

Water enters the drainage basin system as precipitation. This may be in any form, such as snow or rain. Most drainage basins have some vegetation. The precipitation may be caught on the leaves of plants. This is called interception. Generally, it is greatest in summer as this is when deciduous plants have leaves.

From the surface of the plant, the water may evaporate back into the air, or flow down the stem of the plant to reach the ground. At this point the water has moved from the store in the vegetation to be part of the surface storage. If conditions are right, it will then seep into the soil. This process is called infiltration. Soil normally has small pockets of air called pores, which allow the water to get into it. Once in the soil, gravity will pull the water downwards and it will move down through the soil as throughflow, until it reaches the water table, where all the pores in the soil, or rock, are already full of water, so it cannot move any further downwards. Instead it now flows laterally (sideways) into the nearest river as groundwater flow.

Any water that hits an impermeable surface, with no pores, such as tarmac, cannot infiltrate the soil below. It simply flows over the surface as surface runoff (also called overland flow) into the nearest river.

Although some precipitation can fall directly into the river, most water reaches a river by a combination of surface runoff, throughflow and groundwater flow. It takes water the longest time after falling to reach the river by groundwater flow, since it has had to flow through so many stores to get to the river channel.

Water may leave the drainage basin through the river by going out to seas, lakes or oceans. The rest is either stored within the basin or may be evaporated from the ground, vegetation or soil. Plants respire some of the soil moisture out through their leaves, a process called transpiration.

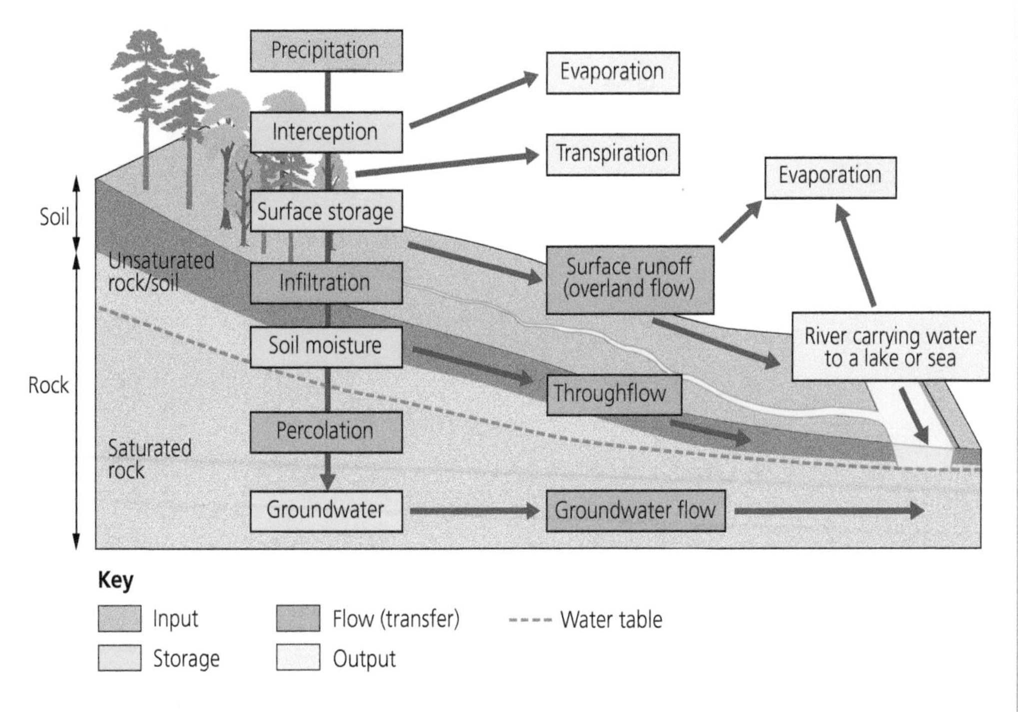

▲ **Figure 4** The drainage basin system.

Activities 

1. Draw an unlabelled diagram of a drainage basin. Swap with a partner and see if they can fill in the missing labels.
2. After researching river sources, name three types of river source.
3. Describe the process of infiltration and suggest two factors which might influence the rate of infiltration.
4. Name and describe the slowest transfer process of water into a river.
5. Explain why rivers are more likely to flood after a long period of rainfall.

By the end of this section you will:

- understand how gradient, depth, width, discharge and load change along the long profile of a river and its valley.

Changes along the long profile of a river

What changes take place along the long profile of a river?

Rivers can be divided into upper, middle and lower courses. To investigate how a river can change downstream, it is possible to examine a local river like the Glendun River in County Antrim. The location of this river is shown in Figure 16 on page 18.

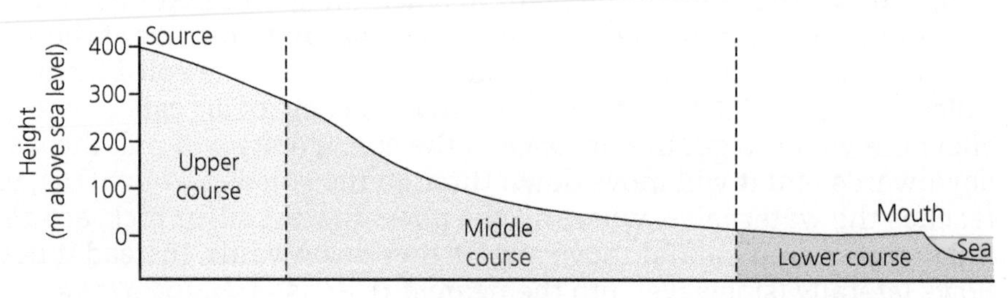

Expected characteristics	Upper course	Middle course	Lower course
	Narrow channel Shallow channel Vertical erosion Hydraulic action, abrasion and attrition Transport of load by saltation, some traction at high flow Large sized, angular load.	Channel widens due to lateral erosion Less hydraulic action All methods of transport can be seen, but suspension is most common Erosion and deposition seen on meanders Load becomes smaller and more rounded.	Channel is deepest and widest Some lateral erosion Deposition more common than erosion Load moved mostly by suspension Load is now small in size, mostly sand, silt and clay.

Cross section

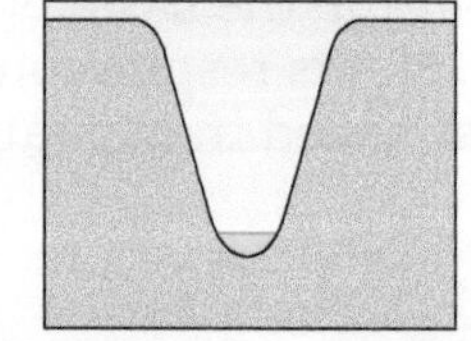

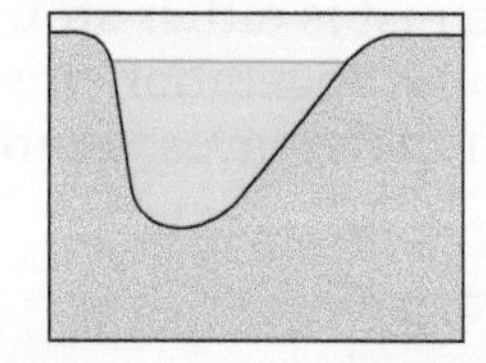

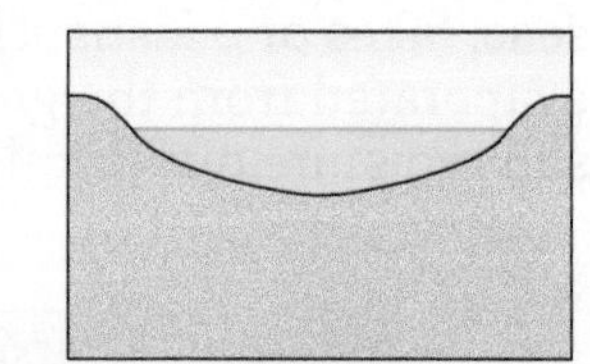

▶ **Figure 5** The typical long profile of a river.

The Bradshaw Model is another way of showing how a river's characteristics change upstream and downstream, given natural conditions. The wider the triangle, the more there is of that particular item. So, for example, the depth of the river channel is typically widest and deepest downstream, in the lower course, towards the mouth of the river. Of course, not all rivers follow these models.

Upstream **Downstream**

Channel depth

Occupied channel width

Mean velocity

Discharge

Volume of load

Load particle size

Channel bed roughness

Gradient

▲ **Figure 6** The Bradshaw Model.

Activities

1. State two river characteristics that typically increase downstream.
2. State two river characteristics that typically get smaller downstream.

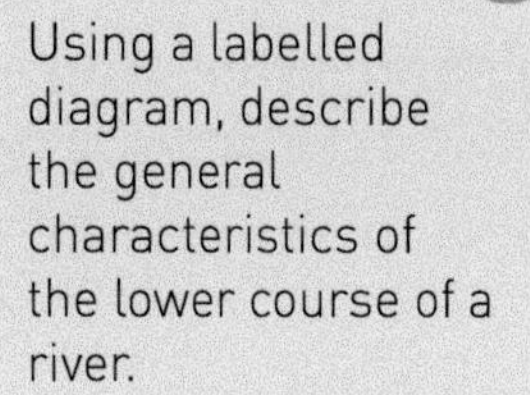

Skills activity

Using a labelled diagram, describe the general characteristics of the lower course of a river.

How do a river's gradient, depth, width, discharge and load change?

Measuring the Glendun River

Various fluvial (river) characteristics are measured at regular points (every 1 km) along the Glendun River. This type of sampling is called systematic sampling and it allows the investigation of continuous changes as distance increases from the source of the river.

A group of pupils investigated this river. Figure 8 shows what the students measured.

▼ **Figure 7** Measurements being taken on the Glendun River.

Gradient

Gradient is a measure of the slope of the land. On a steep slope the gradient is high and on flat land the gradient is low. A river's gradient is measured using a clinometer and ranging poles over a set stretch of the river, such as 5 or 10 metres. Ranging poles mark out the stretch of river. The clinometer is placed against the height mark on the upstream pole and the angle is observed to the equivalent height mark on the downstream pole.

Depth

This is measured using a metre stick. The stick is lowered into the water every 10 cm, and the distance from the top of the water to the river bed gives the depth of water. An average of all these readings is taken.

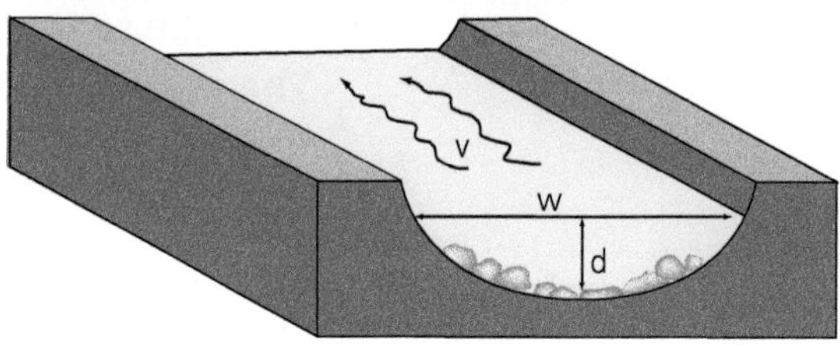

w = width v = velocity d = depth

width × depth = cross-sectional area
velocity × cross-sectional area = discharge area

▲ **Figure 8** Aspects of a river that can be measured.

Width

This is measured by placing one end of a measuring tape at one side of the river channel, then pulling it out to the other side of the channel. The distance is the width of the river.

Discharge

Discharge is the amount of water passing any point in a river in a certain time, normally given as cubic metres of water per second (cumecs). It is calculated by multiplying the cross-sectional area of a river channel at a certain point by the speed (velocity) of the river at the same point.

The cross-sectional area is obtained by multiplying the width of the river by the average depth. The speed (velocity) of the river is recorded using a flow metre that when dipped into the river gives a digital reading of the speed of flow in metres per second.

Load

The load of a river is the material it is carrying, ranging from small sediment to large boulders. Load can be dissolved in the river water (solute load), be carried within the river water (suspended load) or lie on the bed of the river (bedload). It is very hard to measure the size of the load in suspension or solution, so, instead, we can concentrate on the load lying on the channel bed – called bed load. This load is measured for size and roundedness. By measuring the longest axis of 15 random samples at each point an idea of the size of the load is obtained. Each stone is then given a rating for roundedness.

Results of measuring the Glendun River

To help see the overall trends, here is a selection of results obtained from the Glendun river.

▼ **Table 1** Results of measuring the Glendun River.

	Upper course (Station 1: 1.5 km from source)	Middle course (Station 3)	Lower course (Station 6: 16.5 km from source)
Width (m)	2.7	10.4	14.2
Depth (m)	0.14	0.33	0.46
Discharge (cumecs)	0.08	0.2	5.1
Load – long axis (cm)	26	12	7
Load – roundedness	angular	sub-angular	rounded

Read pages 13–14 about erosion, transportation and deposition to help you interpret these results and understand the explanations below.

Going downstream from source to mouth, the Glendun River gets wider. At Station 6 (16.5 km from the source) the river is just over five times the width it is at Station 1, only 1.5 km from the source.

The river gets deeper. At the station in the lower course the river is 32 cm deeper than it is in the upper course. This can be explained by the fact that there is more lateral erosion and vertical erosion occurring downstream from the source.

The wider and deeper river channel size downstream relates well to the pattern of increasing discharge. Because discharge is calculated by multiplying the cross-sectional area of the channel by the river's velocity, then it follows logically that as the cross-sectional area increases, so does the discharge. The river is receiving additional water from the tributaries that are entering it at regular intervals within the Glendun valley: these will also cause the discharge to be greater downstream. In the upper course, by contrast, very few tributaries have contributed to the flow. Finally, the velocity of the river is also greater in the lower course as the water flowing in the river channel does not have to overcome as much friction as that in the upper course, which has angular rocks and a shallow channel.

Most of the weathering of bare rock happens in mountain areas, where it is exposed. This material can then fall down the steep valley sides into the upper course of the river. It is still very angular, as the results show. As it moves downstream it hits the sides of the river bed, and also other rocks that make up the load. This knocks the sharp edges off the material, smoothing its sides and making it rounded. The load of the river, therefore, is noticeably more angular in the upper course, but becomes rounded in the lower course – even on a relatively short river such as the Glendun.

Activities

1 Using graph paper or a computer, draw graphs to show how discharge and load size change along the long profile of the Glendun River.
2 Analyse how discharge and load size change by describing your graphs.
3 Interpret why discharge and load size change by explaining the river processes that have caused the changes.
4 Suggest suitable health and safety measures a group should take when doing a river study.

2 River processes and landforms

River processes

By the end of this section you will:

- understand how rivers erode, transport and deposit material.

How does erosion work?

All rivers contain minerals and solid material: this is known as the **'load'** of the river. When rivers have a large load made up of coarse materials these scrape or rub against the channel bed, eventually lowering the level of the bed, to create steep valley sides. This is called vertical (downward) erosion.

In sections of the river channel where the river is flowing particularly fast, the water has enough energy to wash away part of the bank of the river. This can lead to undercutting and collapse. As this is a sideways motion, it is called lateral erosion.

There are four main types of erosion:

Attrition is the collision of rock fragments (suspended in water) against one another. Rock particles are broken into smaller pieces and become smoother as the process continues.

Abrasion or corrasion is the grinding of rock fragments, carried by a river, against the bed and banks of that river. This action causes the river channel to widen and deepen. This grinding is most powerful during a flood, when large fragments of rock are carried along the river bed.

Hydraulic action is a form of mechanical weathering that is caused by the force of moving water. It can undermine riverbanks on the outside of a meander or force air into cracks within exposed rock in waterfalls such as Niagara (see Figure 11).

Solution or corrosion is the process by which river water chemically reacts with soluble minerals in rocks and dissolves them.

How does transportation work?

Weathered material falling into a river from the valley sides forms 90 per cent of the river's load. The remaining 10 per cent is the result of erosion caused by the river on its own banks and bed.

Rivers move their load through transportation in four ways as shown in Figure 9 (solution, saltation, suspension and traction).

Solution – soluble minerals dissolve in the water and are carried in solution. This is a chemical change affecting rocks such as limestone and chalk, and may result in discoloured water. For example, the rivers of the Mournes often appear yellow/brown as they are stained from iron coming off the surrounding peat bog.

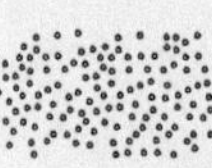

Suspension – the smallest load, like fine sand and clay, is held up continually within the river water. This makes the water appear opaque. Some rivers carry huge quantities of suspended material.

Direction of flow

Traction – the rolling of large rocks along the river bed. This requires a lot of energy. Load carried in this way is called bed load. The largest bed load will only be moved like this in times of severe flood.

Saltation – the bouncing of medium-sized load, like small pebbles and stones, along the river bed.

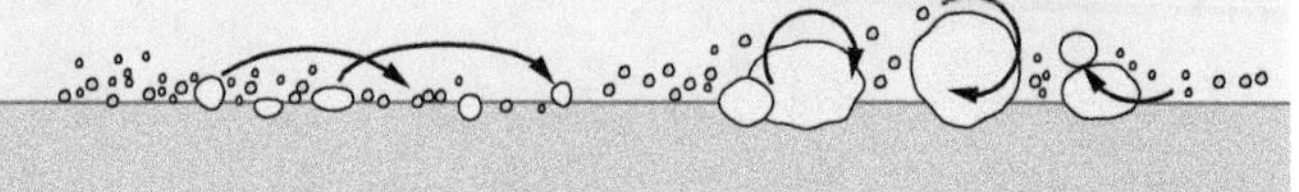

River bed

Figure 9 Methods that a river uses to move its load.

Try using a mnemonic or a rhyme to help you remember all the different types of erosion and transportation.

How does deposition work?

When the velocity of the river is reduced, the energy of the water decreases, and so the water can no longer erode or transport material. Instead, the load is dropped, starting with the largest (and heaviest) particles. This process is called deposition.

Conditions when deposition is likely to occur are shown in Figure 10.

It is the combination of erosion, transportation and deposition that creates the general landforms seen along a river channel.

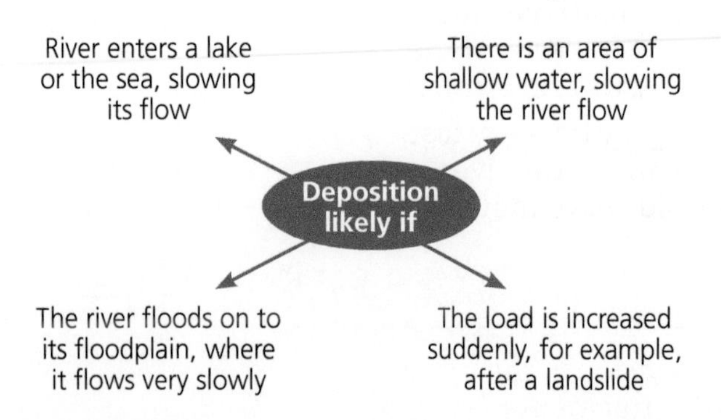

Figure 10 Conditions that are likely to result in deposition.

Activities

1. Make double-sided flash cards to explain the ways a river carries out the following processes. Use these with a partner in class to see if you can remember them all.
   a. erosion
   b. transportation
2. Explain why deposition is more likely to happen when the velocity (speed) of a river slows.
3. State three differences in the conditions of a river that lead to erosion or deposition occurring.

Figure 11 Niagara Falls.

River landforms

By the end of this section you will:

- understand how waterfalls, meanders and slip-off slopes form.

Waterfalls

Within the drainage basin, waterfalls are generally found in the upper course of a river. This is near the river's source area where the landscape is still quite mountainous.

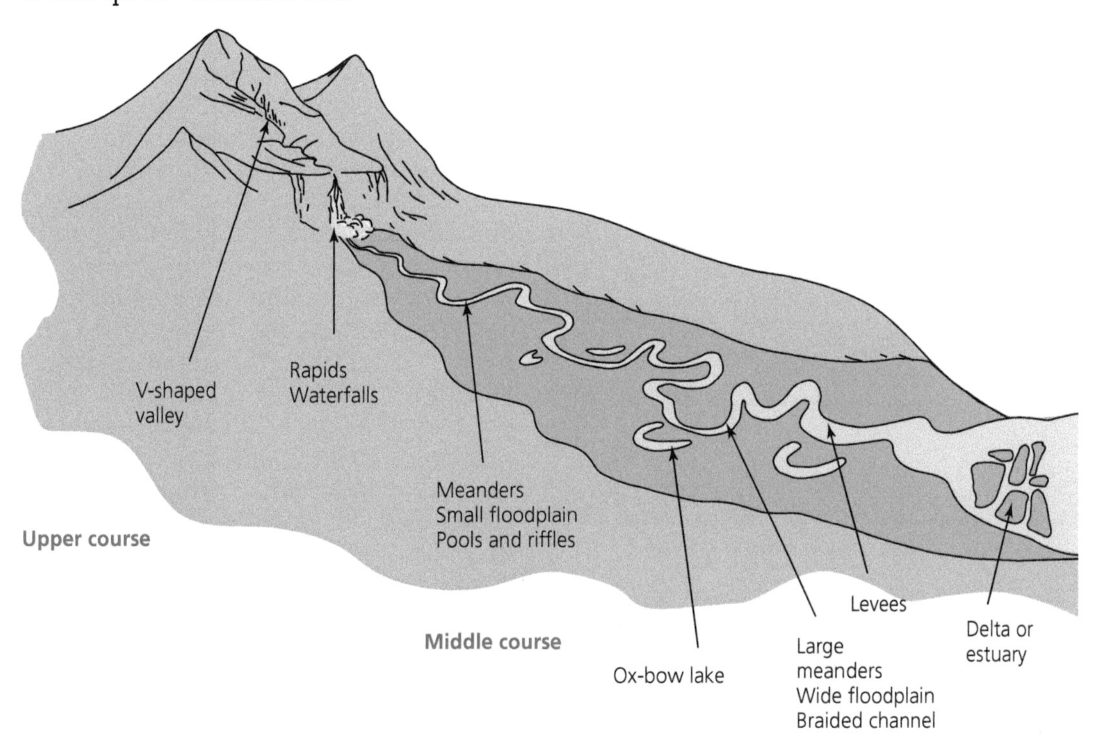

▲ **Figure 12** Landscape features in a drainage basin.

Waterfalls form where there are alternating layers of hard rock and soft rock. As the river passes over the exposed, less resistant soft rock on a river bed, it is able to erode it at a faster rate than the harder rock, so a step in the river bed develops. The force of hydraulic action and abrasion deepens this step until a waterfall is formed. Eventually, erosion makes a deep pool under the waterfall, called a plunge pool, and the hard rock will begin to hang over this pool. When it becomes too unstable, the hard rock overhang will collapse and the waterfall will retreat backwards, leaving a gorge.

One of the most famous waterfalls is Niagara, where hard limestone lies over softer shale (see Figure 11 on page 14).

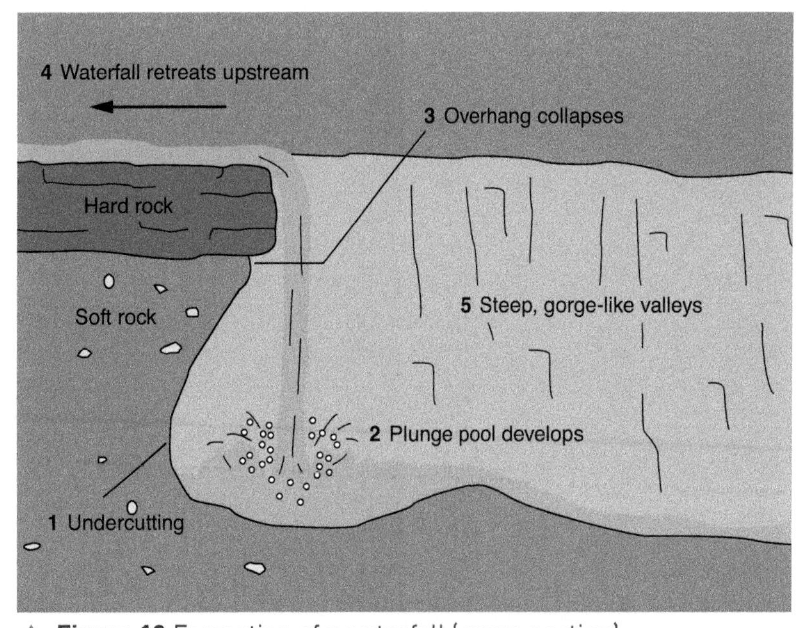

▲ **Figure 13** Formation of a waterfall (cross-section).

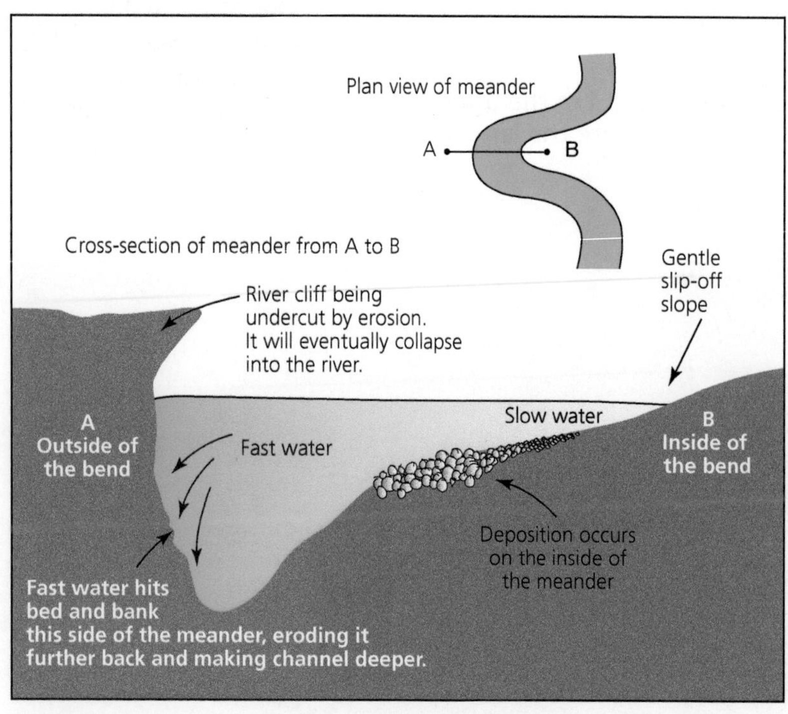

Figure 14 Plan and cross-section of a meander.

Meanders

Meanders are bends that develop in a river channel as the gradient (slope) of the river evens out. They are continuously-changing features that are the result of differences in the velocity of the water in the river across its channel. Where water flows fastest in the channel it spirals downwards, causing vertical erosion, deepening the river channel and creating a river cliff on the bank. Opposite this, water flows very slowly and does not have enough energy to erode the bank. This slowly-flowing water cannot hold up the load it is carrying, so deposits the heaviest material first, then the next heaviest, and so on, until only the smallest clay particles may be left in suspension. This leads to a lop-sided cross-section through a meander, as shown in Figure 14. Over time, this area builds up to create a clear slip-off slope.

In the middle and lower courses of a river, meanders are constantly being formed and reformed. The bends can get bigger and sometimes they can even be cut off altogether – this is how an ox-bow lake is formed.

Floodplains and levees

When a river overflows its banks the velocity, and therefore energy, of the water quickly decreases: the river must deposit much of its load on to the floodplain. The largest load is deposited first. Over time, after repeated floods, layers of deposited material build up to form natural embankments called levees. During low flow, the river may deposit material on its bed as the amount of water in the river falls. This could result in the river drying up completely. If load is deposited on to the river bed, and not washed away later in the season, the river bed can be raised. In some cases the river may end up flowing above the level of the floodplain.

Levees can be artificially strengthened and raised to protect a floodplain from flooding.

Activities

1. Draw a field sketch of Niagara Falls using Figure 11. Ensure you label it fully to explain how it formed.
2. On a blank diagram of a meander in cross-section, add the following labels: deep water, shallow water, erosion, deposition, slip-off slope, river cliff and point bar.
3. Write five statements to summarise the text on meanders.

Identifying river landforms and land uses

By the end of this section you will be able to:

- interpret aerial photographs and OS maps to identify river landforms and land uses.

How do aerial photographs help identify landforms?

Aerial photographs are photographs that have been taken from above, usually from a plane, a helicopter or a drone. They can be useful to examine large (macro) features of a landscape. If taken over a period of time, they can document landscape change, such as the formation of an ox-bow lake or land use change (such as regeneration) in cities.

To use aerial images properly they must have a clear title, a six-figure grid reference (or GIS map to show location) and direction of view. Normally such images are carefully labelled (annotated) to explain the notable features.

Figure 15 Aerial photographs of the River Thames in London (above) and the Usumacinta river in Guatemala (right).

Tip 

You need to have basic map skills to cover this section. Look at the Ordnance Survey website to refresh them.

Activity 

Complete a field sketch of the aerial photograph of the Usumacinta river adding in labels for erosion, deposition, point bar and river cliff.

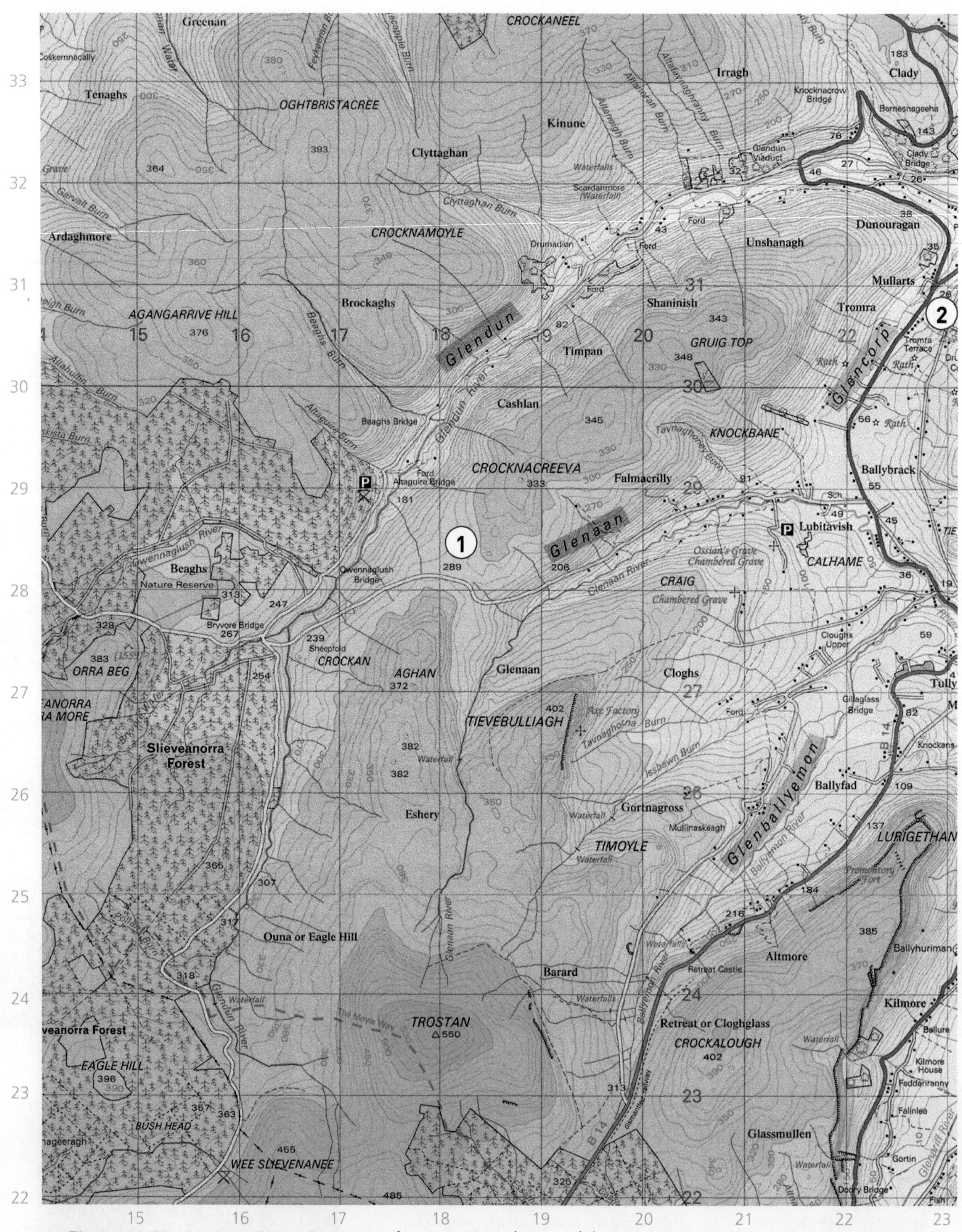

Figure 16 The Glendun River, Co. Antrim (scale 1:50,000). Note: (1) shows the end of the upper course of the river and (2) shows the end of the middle course.

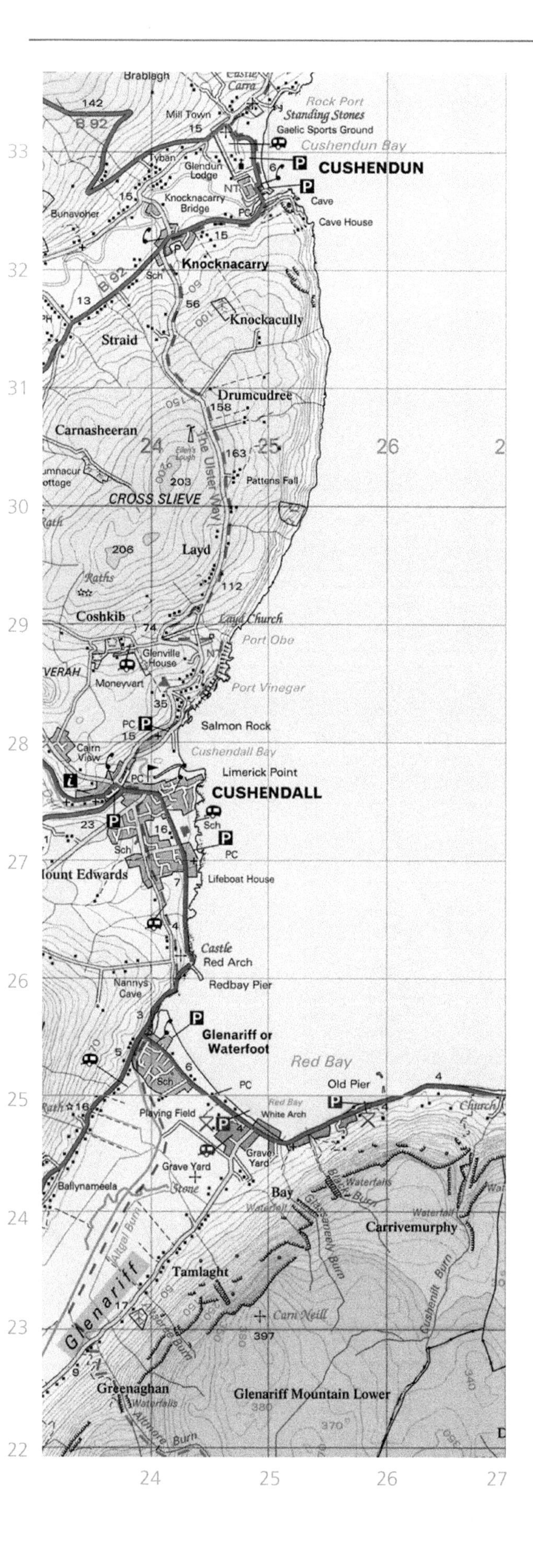

How do Ordnance Survey (OS) maps help identify landforms?

River landforms can be seen on OS maps. If you are drawing a sketch map of a river from an OS map, the sketch should have a clear title, numbered grid squares and annotation to explain the notable features.

In Figure 16 you will see an OS map extract of the area around the Glendun River.

Using OS maps to identify river landforms and land uses

Activities

1. What evidence can you find on the Ordnance Survey (OS) map to support the view that tourists visit this area? State your evidence from the map, draw the relevant OS map symbols and give the six-figure grid references.
2. Calculate the distance the river covers in its middle course. Give your answer in kilometres. (Tip: be sure to check the scale of this map. How many centimetres on the map represent a kilometre on the ground?)
3. In the upper course of the Glendun River there is a waterfall. Give its four-figure grid reference.
4. How would you describe the valley sides of the Glendun in grid square 1830? What is the evidence from the OS map to support your opinion?
5. Why would much of the land in this area be of limited use to farmers? How could farmers make best use of the land?
6. How would you describe Slieveanorra Forest?
7. How would you describe the coastline south of Cushendun?
8. Using the OS map extract on page 18 draw your own annotated sketch map to show the course of the Glendun River. Your map should show:
   a. the course of the Glendun River
   b. the Glendun valley
   c. major tributaries
   d. the upper, middle and lower courses of the river
   e. wide floodplain
   f. land over 300 m above sea level
   g. Slieveanorra Forest
   h. roads
   i. Knocknacarry and Cushendun
   j. a waterfall
9. Devise a suitable key for your sketch map.
10. Give your sketch map an appropriate title.

By the end of this section you will be able to:

- understand the physical and human causes of flooding.

What causes rivers to flood?

Rivers flood due to a mix of human (where human activity has changed the landscape) and physical (where changes have occurred naturally) factors. Flash floods are generally caused by sudden instances of heavy rainfall, which are typically associated with an extreme weather event such as a hurricane or tropical storm, or sometimes meltwater from ice.

Physical causes of flooding

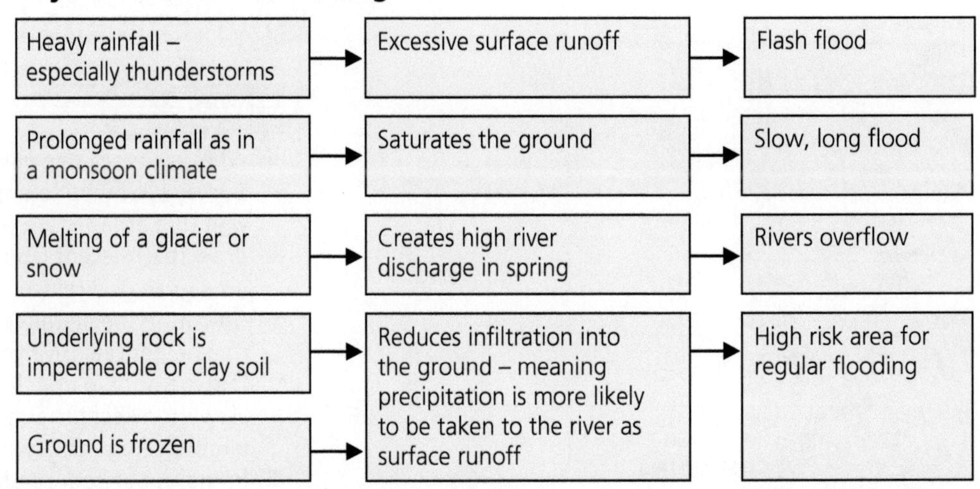

Human causes of flooding

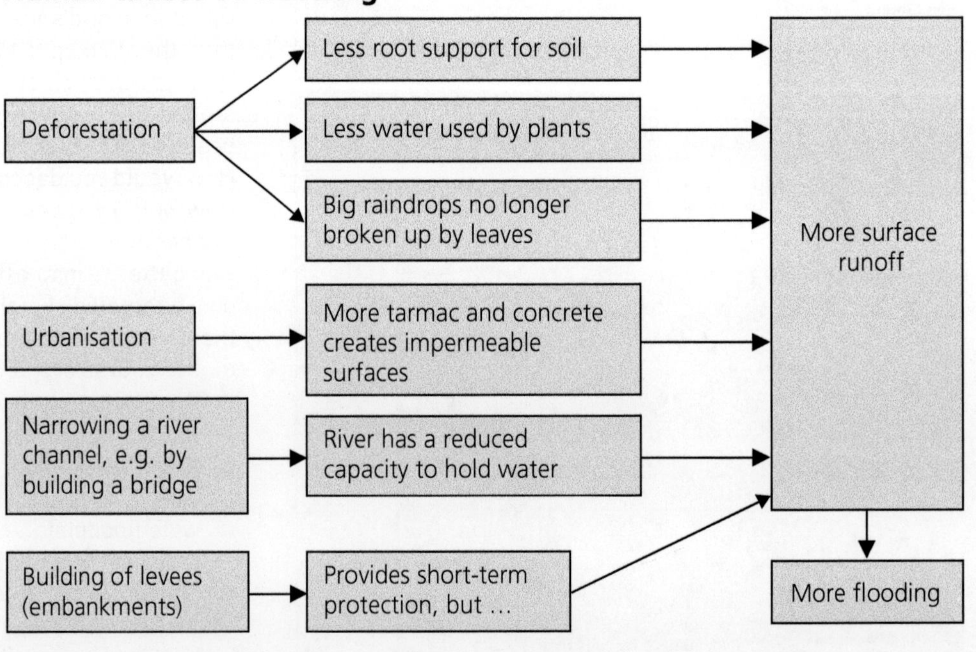

Figure 17 Physical and human causes of flooding.

Tip

You need to know which causes of flooding are physical and which are human, so create a jumbled-up list then sort them out. Either cut them out and move them around or try colour-coding them.

Activities

1. Briefly explain why rivers flood, using the terms 'physical' and 'human' within your answer.
2. Describe two physical and two human causes of flooding in one area in the British Isles you have studied.

CASE STUDY

Somerset Levels, 2014

- The Somerset Levels is a low-lying region in the South West of England. As it has a naturally high water table and poor drainage it is an area prone to flooding. However, the winter of 2013–2014 saw it experience prolonged flooding which was described as the worst in over a century. Villages were cut off as roads became impassible and animals had to be moved from some of the 6880 hectares of flooded farmland.

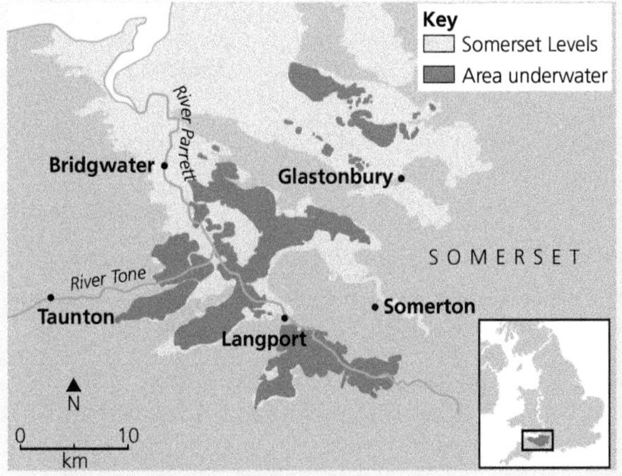

Figure 18 Map of Somerset showing the Levels and flooding, and a photo of the flooding.

Causes of the flooding

Physical causes

- The combination of impermeable bed rock (the clay) and low interception levels (the grass) means that the area is naturally at risk of flooding. Southern England received 207mm of rainfall in January 2014 alone, this was more than twice the expected total for that month.
- There was a series of severe winter storms in December and January in southern England during 2013–14. December 2013 was one of the stormiest Decembers on record, and one of the windiest months since January 1993 – figures released by the Met Office on 30 January indicated that southern England and parts of the Midlands had experienced their highest January rainfall since records began in 1910.
- High tides cause floodwater to back up along the rivers across the Levels and Moors. This was exacerbated by the river levels being higher than usual because of the recent rains, and because they had not been dredged (see the human causes in the next column).

Human causes

- The Rivers Tone and Parrett had not been dredged properly in 20 years, leaving farmland and homes without proper defence from the floods. This resulted in hectares of land being left underwater from the storms which began in December. Locals believe if the rivers had been dredged, this would have cleared them of silt, making them wider, deeper and easier to maintain. It would also have created more capacity to carry away flood waters, draining the floodplain far more quickly.
- In addition, extra water was also sent into the Levels from Taunton and Bridgwater, as part of a scheme where water is pumped away from these areas to protect new homes built on former floodplains.

Weblink 

Go to www.slideshare.net/juicygeography/flooding-on-the-somerset-levels to find out more about this topic.

By the end of this section you will be able to:

- recognise the impacts of flooding on people
- recognise the impacts of flooding on the environment.

Tip 

Make sure you learn a balanced number of impacts on people and the environment.

The impacts of flooding

What are the impacts of flooding on people and the environment?

Impacts of flooding on people

Positive impacts	Negative impacts
Replenishes drinking water supplies, especially wells. Provides sediment (other terms are silt or alluvium) that naturally fertilises the soils of the floodplain. Countries such as Bangladesh and Egypt rely on floods to help crops like rice grow. Can encourage innovative solutions in future building design, such as building houses on stilts to avoid floodwater, or tiling all ground floor walls and floors to facilitate easy clean-up after flooding.	Spreads waterborne diseases such as cholera. People and animals can be made homeless or even drown. Buildings and infrastructure (roads and railways) can be damaged or destroyed. Crops grown on fertile floodplains can be washed away in a flash flood. Can increase house insurance costs for householders, or even make it impossible for them to insure their home at all.

Impacts of flooding on the environment

Positive impacts	Negative impacts
Fish benefit as they can breed in the standing floodwater. In dry areas, floods bring relief from drought, providing drinking water for wild animals.	Flooding can wash chemicals or sewage into the local rivers and so pollute them. Wild animals may drown or lose their habitat during a flood.

Activities

Choose one of the following two statements and write an argument to support it. Use the table above to help you.

1. Allowing rivers to flood is important.
2. Stopping rivers from flooding is important.

Figure 19 One of the impacts of flooding.

Methods used for managing rivers

By the end of this section you will be able to:

- recognise the difference between hard and soft flood management methods
- understand how hard engineering methods work
- understand how soft engineering methods work.

What are the methods used for managing rivers and floods?

Flood management schemes aim to control rivers and reduce the risk of unwanted flooding. Planners can respond to flood hazards by changing the river through hard or soft engineering methods.

In Northern Ireland, the Department of Infrastructure looks after rivers through the Rivers Agency. The aim of the agency is to reduce the risk to life and property from flooding from rivers and the sea, and to manage rivers and the coast in a sustainable manner.

What are hard engineering methods?

Hard engineering methods often involve using large artificial structures to control the river, breaking its natural cycle of flood and subsidence. However, these measures are not sustainable in the long term. They may involve:

- Building a dam or reservoir in the upper course, which can then be used for leisure and hydroelectricity, but can also flood farmland, displace local people and destroy habitats.
- Deepening and widening the river channel in order to increase its cross-sectional area, allowing it to contain more water. Therefore the discharge has to be much greater to create a flood.
- Straightening a river by cutting off the loop of a meander to speed up the flow of a river and therefore reduce flood risk.
- Building high embankments along the sides of the river to contain any floodwater. This might be done by simply raising natural levees.
- Building a flood wall along the course of a river where it flows through settlements. These flood walls are expensive and do not look natural, but they can be effective if they are maintained properly and built high enough.
- Creating storage areas, which are a bit like temporary lakes. As rivers rise, extra water can be pumped or diverted into them, then emptied back into the main river when the risk of flooding is over.

▲ **Figure 20** Spelga Dam in County Down.

Find out about river management in your local area from the Rivers Agency website: www.infrastructure-ni.gov.uk/topics/rivers-and-flooding.

What are soft engineering methods?

Soft engineering methods are generally sympathetic to the natural landscape, so they tend not to damage the river for future generations. They are more sustainable than hard engineering methods and may involve:

- Planting trees (afforestation) in the upper course of the river in order to increase interception and reduce the risk of flash flooding. As trees absorb water from deep within the soil, this also reduces the flow of water to the river from soil moisture. This in turn reduces percolation and so also the water within the groundwater store.
- Land use zoning – this is when areas most likely to be flooded are protected from urban development (see Figure 21). High-risk flood areas might be left as pasture for grazing while low-risk zones might become playing fields or parkland. Housing and important buildings like hospitals are only built on the land least at risk from flooding.
- Washlands – these are parts of the river floodplain in the lower course into which the river can flood temporarily. They are a kind of flood storage area. Washlands can prevent the flooding of more valuable land further downstream.

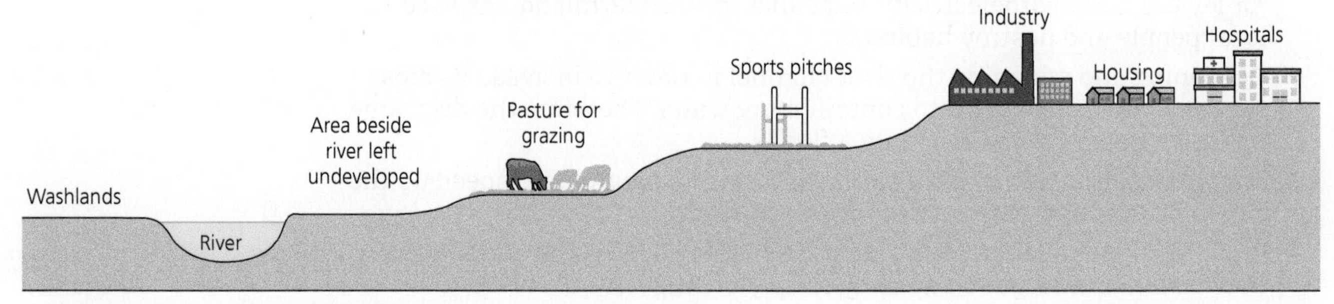

Figure 21 Floodplain zoning.

Activities

1. For each flood control strategy write out five key words or phrases you would include when describing and explaining it, for example: dam, hard engineering, upper course, wall, reservoir, hydroelectricity.
2. Suggest why environmentalists are more in favour of soft engineering methods of flood control than hard engineering methods.

CASE STUDY

The Mississippi River: management strategies

By the end of this section you will be able to:

- investigate a river outside the British Isles
- evaluate the river management strategies used with this river.

Why does the Mississippi River make a good case study?

- The Mississippi River has one of the largest drainage basins in North America – it drains water from a third of the USA and part of Canada. It is located in the south-east of the USA. The river is important as a shipping channel, for recreation, as a supply of hydroelectric power and as a store of drinking water.
- When it flooded in 2011, around 25,000 people were evacuated and the property damage costs were estimated at $3 billion.

Figure 22 Aerial photo of the Mississippi.

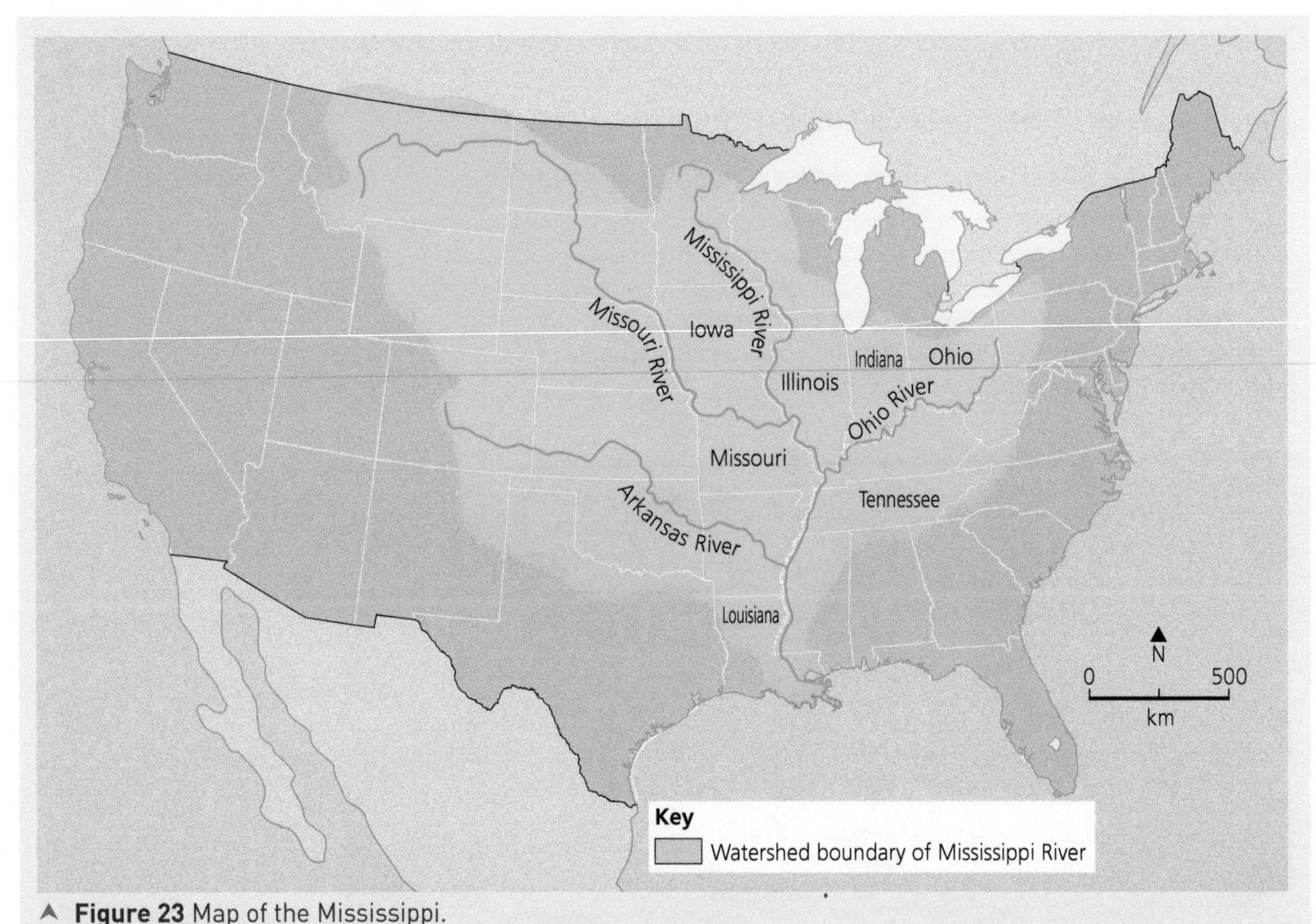

Figure 23 Map of the Mississippi.

What were the hard and soft management strategies in response to the flooding?

Hard engineering methods	Soft engineering methods
▶ Raised levees – levees were raised to 15 m and strengthened to enclose the river channel for a stretch of 3,000 km. ▶ Straightening the river channel – meanders were cut through over a stretch of 1,750 km, creating a fast-flowing straight river channel. ▶ Dams – the flow of the main tributaries, e.g. Ohio River, has been controlled by 100 dams.	▶ Afforestation in the upper course – trees have been planted in areas such as the Tennessee Valley to intercept some of the rainfall and stabilise the soil. ▶ Safe flood zones – building has been restricted in many of the floodplain areas. Also, in areas like Rock Island, where housing had already been built on the floodplain, the housing has been bought by the county and demolished. ▶ Washlands – in 2011, the Morganza Spillway was opened to flood around 12,000 km^2 of farmland in Louisiana deliberately, preventing that water from reaching the city of New Orleans.

Tip 

Remember that to manage a river sustainably, the needs of the present generation must be met without endangering the ability of future generations to meet their needs.

Is the Mississippi River being managed sustainably?

- The Mississippi River is very important to the USA since 18 million people rely on it for their water supply. Therefore carefully co-ordinated management decisions need to be made as each decision in one state may affect millions of people's water supply in another state.
- Current hard engineering methods have proven to be neither totally effective nor sustainable. The river still floods, and indeed the dangerous flash-flood nature of recent floods has been partly blamed on the artificial levees failing. Also, as the river bed silts up alongside the levees, river beds rise and the floodplain ends up below the river level, e.g. in New Orleans, some areas are 4.3 m below the river level.
- For current and future generations, the lack of silt reaching the land means that fertility of the soil is no longer being naturally completed during the deposition of alluvium in the floods. Eventually more and more artificial fertilisers will have to be added to the soil.
- For wildlife, the draining of wetland and lack of silt to maintain the delta are destroying valuable habitats. In the last 75 years, Illinois, Indiana, Iowa, Missouri and Ohio have each lost more than 85 per cent of their wetlands. These wetlands worked like a giant sponge by absorbing rainwater and then releasing it slowly into nearby streams or the groundwater store.
- The soft engineering strategy of afforestation has not been applied over a wide enough area and it takes too long for the trees to become large enough to make a noticeable difference to runoff for the current generation of Mississippi residents. Also, despite planning controls in some states, other states have been developing housing on floodplain land. For example, more homes and businesses have been built in the Mississippi floodplain in the St. Louis metropolitan area since the devastating flood of 1993 than in all the time before then.
- The use of washlands should only be an emergency measure as they use up large areas of space near cities. The 2011 flooding showed there is a lack of such washlands, as in response to the emergency the Army Corps of Engineers took the decision to explode the levee at Birds Point to create a makeshift washland that destroyed dozens of farmsteads with their crops.
- In conclusion, it seems that the management of the Mississippi is currently not sustainable, especially given the background of climate change that is causing higher-than-average rainfall levels in the drainage basin. Re-creating wetlands and re-activating floodplains in strategic locations will result in a more effective and sustainable flood protection system. US researchers have estimated that the restoration of 13 million acres (over 5 million hectares) of wetlands in the upper portion of the Mississippi-Missouri watershed, at a cost of $2–3 billion, would have absorbed enough floodwater to have stopped the 2011 flood – which cost $2 billion in damage.

Activities 

1. Using Figure 23, describe the extent of the Mississippi river's drainage basin.
2. Explain why controlling the Mississippi is important.
3. Create a table to summarise the main soft and hard engineering strategies used to control flooding of the Mississippi.
4. Evaluate the flood management strategies used on the Mississippi – ensure you use the word 'sustainable' within your answer.

Tip

If you are asked to 'evaluate', make sure you include information on at least one soft and one hard engineering measure, then finish with a conclusion that makes a decision on the sustainability/effectiveness of the measures. Back up your decision with evidence.

Sample examination questions

Tip

You are being asked for a definition of a key geographical term.

1 State the meaning of the term 'drainage basin'. [2]

2 Using Figure 1, describe the course of the River Severn. [4]

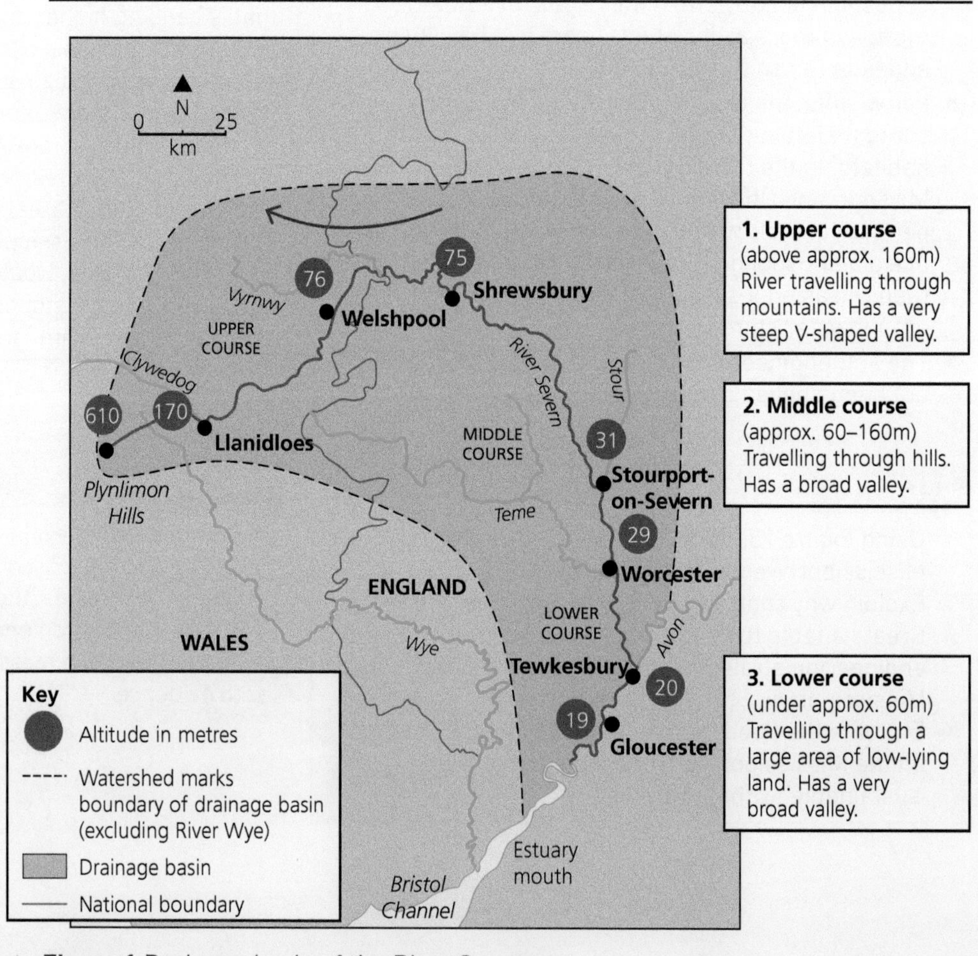

Figure 1 Drainage basin of the River Severn.

3 Copy and complete the following table about changes along the long profile of a river. One has been completed for you. [3]

Increases downstream	River characteristic	Decreases downstream
←	River width	
	River gradient	
	River discharge	
	Size of bed load	

Table 1 Changes along the long profile of a river.

4 Choose one of the river characteristics from Table 1 and explain why it changes. [3]

5 Name two stores within the drainage basin system. [2]

6 Name the process by which water moves horizontally through rock. [1]

7 Figure 2 shows deforestation. Explain how deforestation can influence the flows and stores of water in a drainage basin. [4]

Figure 2 Forestry reduces interception.

8 Using Figure 3, describe the changes in bed load shape along the Glenarm River in Co. Antrim. [4]

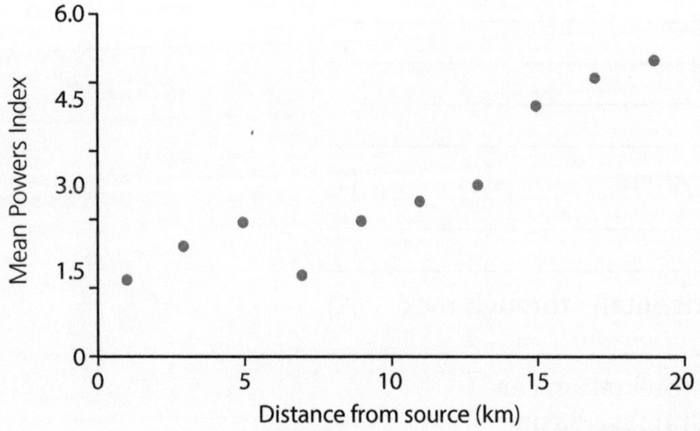

Mean Powers Index

6.0

4.5

3.0

1.5

0

0 5 10 15 20

Distance from source (km)

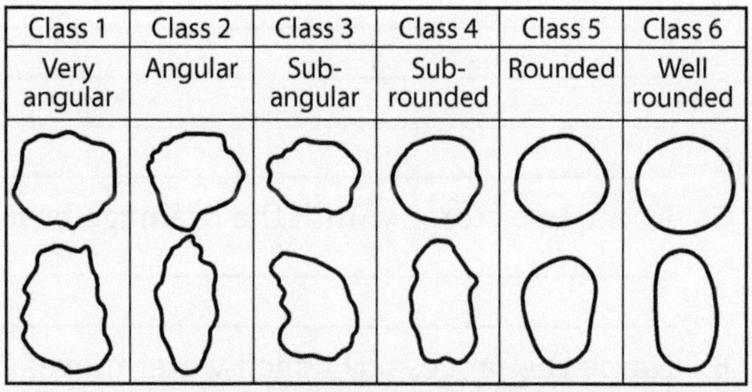

Class 1	Class 2	Class 3	Class 4	Class 5	Class 6
Very angular	Angular	Sub-angular	Sub-rounded	Rounded	Well rounded

The mean of a set of figures is simply the average. In this case it is the average angularity of the bed load in this river. The mean is calculated by adding together all the data recorded at each site on angularity, then dividing that figure by the number of samples taken.

Figure 3 Glenarm River bed load data.

9 Name and describe two erosional processes carried out by a river. [4]

10 Name and describe how a river transports large boulders. [2]

11 Describe and explain the formation of a waterfall. [6]

12 Using Figure 4 to help you, describe and explain the cross-section shape of a river meander.

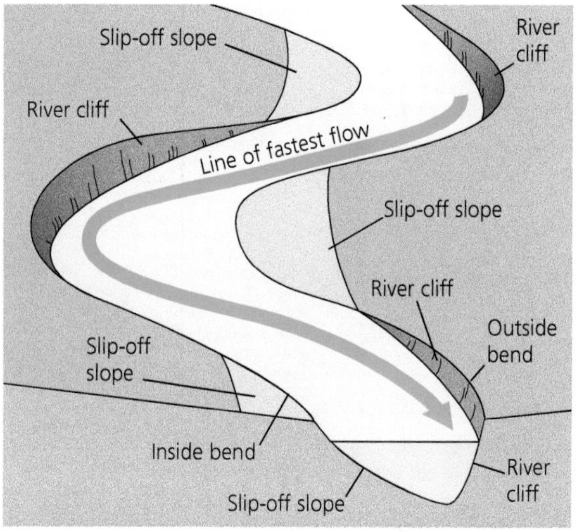

Slip-off slope
River cliff
River cliff
Line of fastest flow
Slip-off slope
River cliff
Outside bend
Slip-off slope
Inside bend
Slip-off slope
River cliff

Figure 4 Cross profile of a meander.

13 Use the map extract in Figure 5 on page 32 and a key of OS symbols to help you answer the following questions.

(i) State the name of the building at GR797178. [2]

(ii) Give the direction from Riversmead farm (804156) to Weir Green (791156). [1]

(iii) The black markings along the river (8016) are levees. Explain how these form naturally. [5]

(iv) Other than the existence of levees, state two pieces of evidence that this map shows part of the Severn's floodplain. [2]

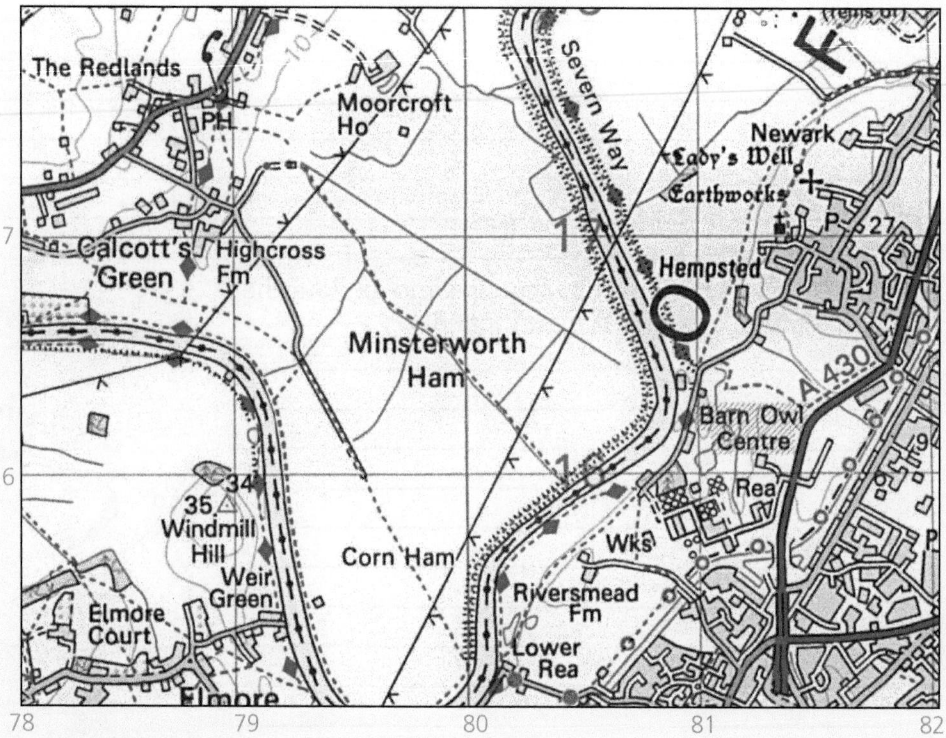

Figure 5 OS map of part of the west side of Gloucester.

14 Explain the causes of flooding for an area in the British Isles that you have studied. [8]

15 Describe two flood control measures used on a river you have studied and evaluate their effectiveness. [9]

UNDERSTANDING OUR NATURAL WORLD

THEME B: Coastal Environments

▲ Coastal groynes.

Why is it important to manage coastlines in some areas and not others?

By the end of this section you will:

- understand the dynamic nature of the coast
- know the difference between constructive and destructive waves.

Constructive and destructive waves

The British Isles gained its 16,000 km long coastline and familiar shape following the end of the Ice Age, when lowland areas filled with meltwater to form the Irish Sea, North Sea and English Channel. The Irish Sea was filled first, cutting Ireland off from mainland Europe, which is why Ireland only has 26 native species of mammals. As Ireland is an island, a stretch of coastline is never more than a few hours' drive away.

Why do coastlines change?

Waves themselves are the main force of coastal change. Waves are caused by wind blowing over a stretch of open water, called a fetch. The greater the fetch, the larger the wave. This is why the Atlantic coastline of Northern Ireland has better surf conditions than the more sheltered eastern coastline bordering the Irish Sea. Although fetch is important, wind speed can greatly affect wave height. The stronger the wind, the bigger the waves.

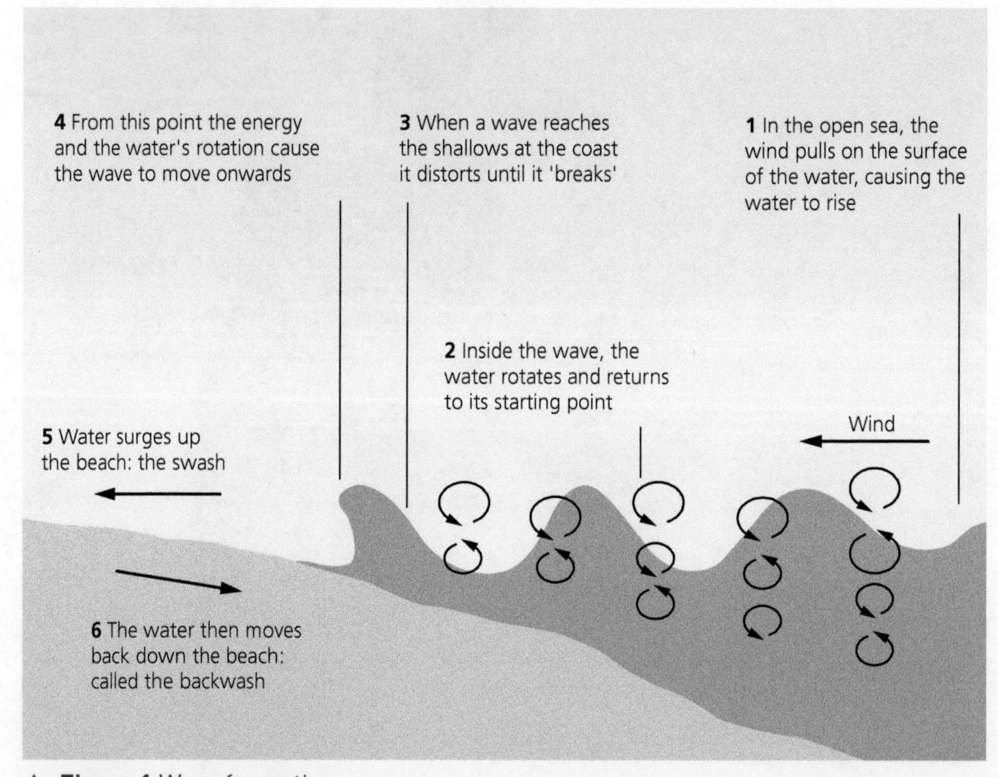

Figure 1 Wave formation.

As a wave approaches shallow water near a coastline, friction caused by the seabed slows down the lower section of the wave more than the upper parts. The upper section of the wave reaches a crest then topples over (breaks) and either hits a cliff face or surges up a beach as the swash of a wave. As the wave retreats, it creates a backwash. Waves with a strong swash and weak backwash are called constructive waves as they push material up a beach. Waves with a strong backwash pull material out to sea, and are therefore destructive waves that erode coasts.

What are the characteristics of destructive waves?

- They have a strong backwash compared to their swash.
- They are high in relation to their length.
- They form frequently (they break at a rate of around fifteen per minute).

Tip

Learn the differences between the two types of wave carefully as they will help you understand how coastal landforms are formed (see pages 37–38).

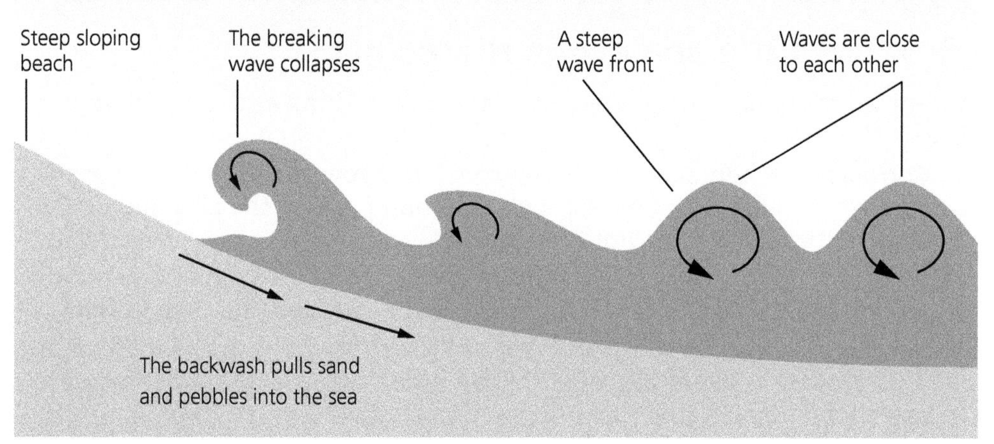

Figure 2 Destructive waves.

What are the characteristics of constructive waves?

- They have a weak backwash compared to their swash.
- They are long in relation to their height.
- They are gentle (they break at a rate of six to nine waves per minute).

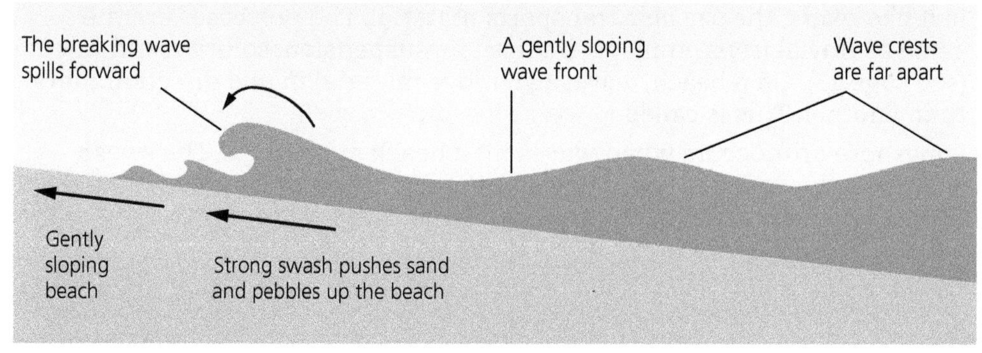

Figure 3 Constructive waves.

Coasts are amazing places that seldom remain the same shape for long. This constant change makes them one of our most dynamic landscapes.

Activities

1. State three ways constructive waves differ from destructive waves.
2. Suggest why it might be unsafe to swim in a coastal area experiencing destructive waves.
3. Compile a list of health and safety rules needed for a group of teenagers conducting coastal fieldwork.

By the end of this section you will:

- understand how the sea can erode, transport and deposit material.

Tip 

Be careful not to mix up coastal processes with river processes; some of the processes are very similar. Remember which section of the examination you are on: is it Section A (rivers) or Section B (coasts)?

Activities 

1. State the differences between coastal erosion and fluvial (river) erosion processes.
2. Describe the process of longshore drift. Use a diagram to help you.
3. Draw a series of diagrams to show how an angular rock would be changed by attrition.
4. Create a set of ten true/false statements about the following three processes carried out by the sea – erosion, transportation and deposition. Record your answers on a separate sheet. Complete a partner's statements and then check your answers.

Coastal processes

As with rivers, the sea also erodes, transports and deposits material. Unlike a river, however, the sea has a much greater force and can move material all around the globe! Coconuts from the Caribbean may be found on the beaches of south-west Ireland and Cornwall, for example.

How does the sea erode material?

When we think of coastal erosion, it is useful to think of the acronym C-A-S-H.

- Corrasion – when a wave hits the coast, it throws sand and pebbles against the cliff face. These knock off small parts of the cliff and cause undercutting. Another word for this is abrasion.
- Attrition – particles being transported by the sea hit against one another, reducing their size and making them more rounded, just like in rivers.
- Solution – seawater can dissolve away the rocks from the seabed or cliffs. This process is especially effective on limestone coasts, and can create spectacular caves. It is also known as corrosion.
- Hydraulic action – the power of the sea can physically wash away soft rocks like boulder clay. Under storm conditions with strong waves, hundreds of tonnes of seawater can hit the coast. Also, air can be trapped in small cracks within a cliff when a wave breaks against it. This compressed air can widen the cracks, eventually leading to sections of cliff breaking away from the main cliff face.

How does the sea transport material?

Just like rivers, the sea also transports material. The processes are the same as fluvial (river) transport: saltation, suspension, solution and traction (see page 13). On a beach, waves can move material in one direction more than another. This is called longshore drift.

Longshore drift occurs when waves hit a beach at an angle. The swash moves up the beach at this angle, but the backwash draws down straight. Over time, material is transported along the coast.

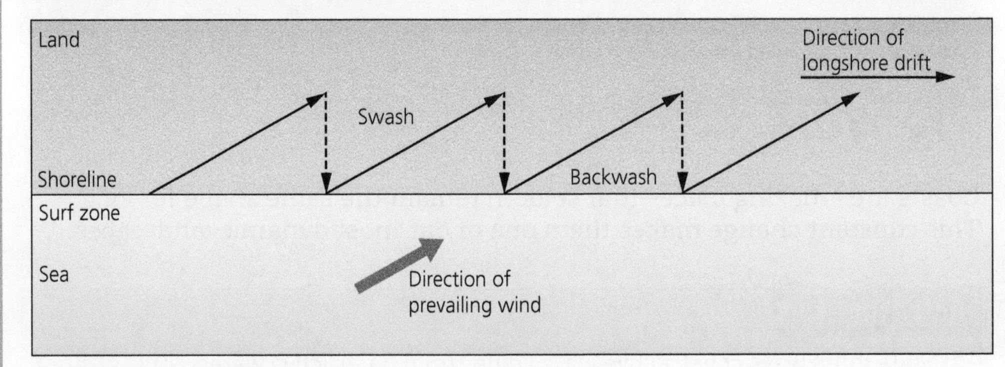

Figure 4 Longshore drift.

How does the sea deposit material?

When the load of the seas and oceans builds up on the coastline it forms beaches, spits and sand dunes. This material is added by constructive waves. Deposition occurs during periods of light winds. This means that the summer is the most common period for this process to occur in the UK. Constructive waves are most effective in sheltered coastal locations such as bays.

Coastal landforms

By the end of this section you will:

- understand the formation of landforms caused by coastal erosion
- understand the formation of landforms caused by coastal deposition.

What coastal landforms are caused by erosion?

Headlands

Stretches of coastline do not erode evenly. Sections of cliffs which are made from hard rock resist the hydraulic and corrosive power of the sea. Eventually they stick out, forming headlands. Headlands face the full force of destructive waves and so are often the location of coastal landforms created by erosion, such as caves, arches and stacks, which are explained later in this section.

Cliffs and wave cut platforms

A cliff is a vertical rock face along the coast. The shape of the cliff is determined by the nature of its geology. The type of rock the cliff is made from determines how resistant it is to erosion, and the way the layers (strata) of the rock are angled can determine the shape of the cliff. Where the rock is hard, tall and awe-inspiring cliffs such as the Cliffs of Moher can form, but with weaker rocks, like the boulder clays of the Holderness coast in England, erosion is faster but the cliffs are less dramatic.

A wave cut platform is the narrow flat area often seen at the base of a cliff. It is caused by erosion. First a notch is formed at the base of the cliff due to corrasion and hydraulic action, which becomes a point of weakness. The upper cliff face is undercut and eventually collapses. This happens again and again, until a new landform, called a wave cut platform, is created at the base of the cliff. It is only fully exposed at low tide.

Figure 5 The Cliffs of Moher.

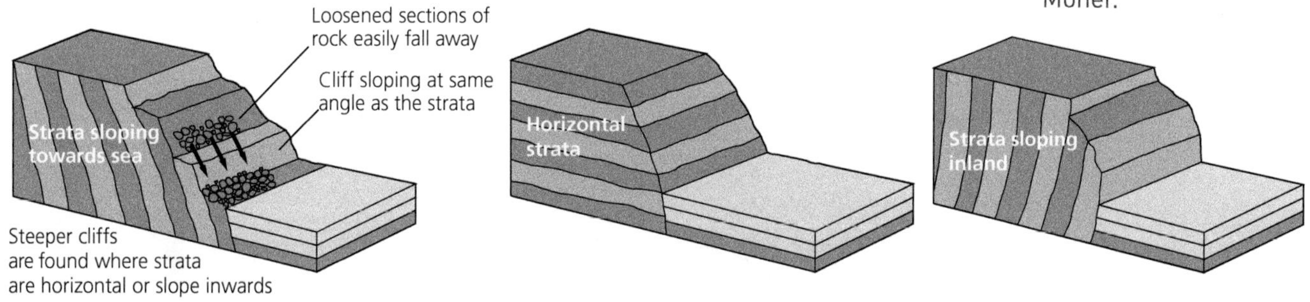

Figure 6 Diagrams showing cliffs shapes formed under various geologies.

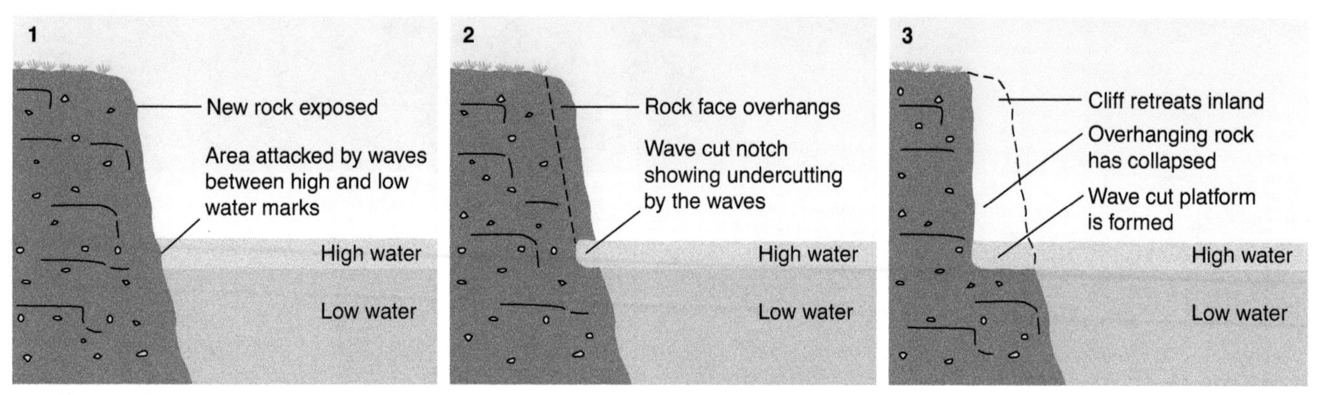

Figure 7 The formation of wave cut platforms.

Activity

Explain the formation of a cliff and wave cut platform.

▲ **Figure 8** A wave cut notch on the Ballintoy coast.

Caves, arches and stacks

A wave cut notch may enlarge into a cave. Following further erosion, the cave erodes through the headland by hydraulic action and corrasion to form an arch. The waves and weathering from the elements undermine the upper portion of the arch until it cannot hold its own weight up, and collapses to leave a stack.

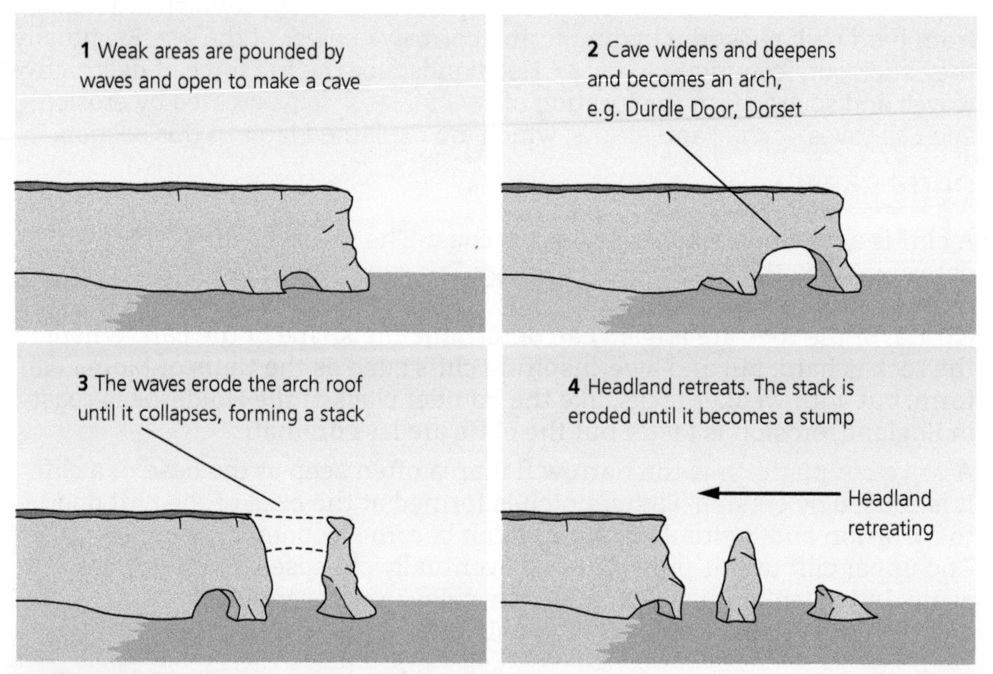

▲ **Figure 9** The erosion of a headland.

➤ **Figure 10** Old Harry Rocks, Studland, Dorset: an example of an eroded headland.

Tip

For each of the coastal landforms learn the sequence of formation as a flow chart. Make sure you are able to provide detail on the type of processes involved in their creation.

Activities

1. State the meaning of the term 'cliff'.
2. Create a flow diagram to help explain the formation of wave cut platforms. Begin with: '1. A wave cut notch is formed at the base of a cliff by corrosion and hydraulic pressure.'
3. Explain how a stack like Old Harry Rocks is formed.
4. Suggest why Dunluce Castle, built on a headland, was considered a good site for a castle.

What coastal landforms are caused by deposition?

Beaches

Beaches are the most familiar coastal landform created by deposition. They are formed in the intertidal area between high and low tide where constructive waves push material such as sand, shingle and pebbles on to the coast. Over time, this material can build up and be blown inshore by wind to create a beach. The supply of beach material depends on erosional rates further up the coast. On sandy beaches, the backwash of the waves continues to remove material, forming a gently sloping beach. On shingle beaches, the energy of a wave is reduced because the large particle size allows percolation, so the backwash is not very powerful, and a steep beach is created.

Activities

1. Why are shingle beaches steeper than sandy beaches?
2. What type of waves form beaches?

Figure 11 A sandy beach – Whitepark Bay, Co. Antrim.

Figure 12 A shingle beach – Bawdsey Beach, Suffolk.

Spits

Spits are depositional features made of sand that look like beaches and extend out from the mainland into the sea. They form if the following conditions are met:

- There is a constant supply of sand or other material from erosion further up the coast.
- Longshore drift operates most of the time.
- The coastline has a sudden change in direction to leave a sheltered bay area.
- The sea is quite shallow.

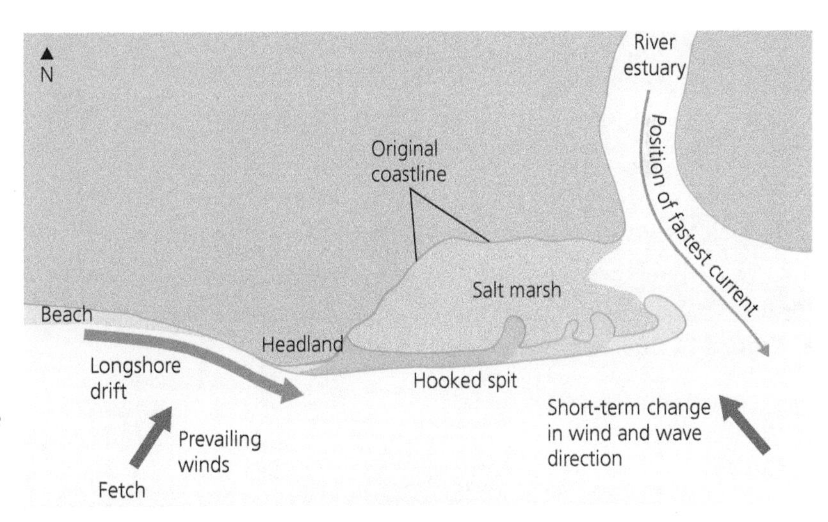

Figure 13 A diagram of a spit.

Spits need a continuous supply of sand and a sheltered bay area to form in. Sand spits occur because waves hit the shore at an angle, moving material along a beach due to longshore drift (see page 36). This happens because although the sand goes up the beach at an angle in the swash, it moves straight back down the beach in the backwash. When this material reaches a natural break in the coastline, the sand continues on and can build up, breaking through the surface and forming a spit attached to the headland. Spits can become stabilised if they occur at an estuary, as silt and mud from the river are deposited behind the spit, making new land. If the angle of the waves changes as the spit extends into the estuary, the tip may turn to create a hooked spit.

One famous example of a spit is Spurn Head in East Yorkshire. The erosion of boulder clay from the Flamborough Head region and longshore drift in a southerly direction have created this landform. It is not fully permanent, as maps of the area show that it has been destroyed by storms four times in the last six hundred years; the most recent was in 1996. Each time it has slowly been rebuilt by longshore drift.

A local example is in Dundrum Bay, County Down, where sand and shingle from Newcastle have built up a small spit. New coastal defences in Newcastle and sea level rises forecast for the region have jeopardised its future survival.

Figure 14 Google Earth image of Spurn Head.

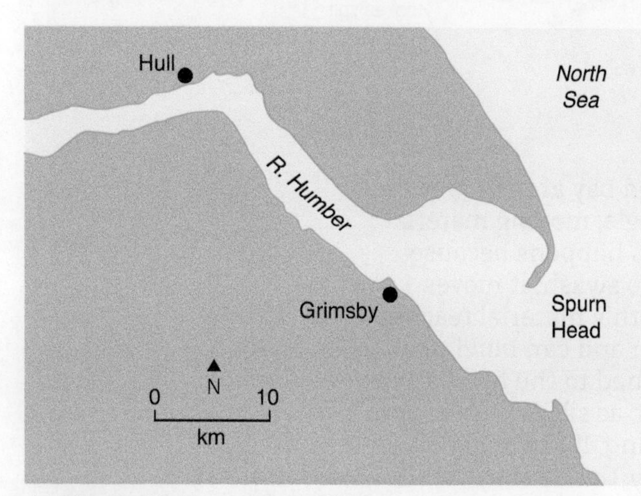

Figure 15 Map of Spurn Head.

Activities

1. Name and describe the coastal process which helped to create Spurn Head.
2. What direction is Spurn Head from Hull?
3. Using the scale on Figure 15, state the approximate distance between Grimsby and Hull.
4. Look at the satellite image of Spurn Head in Figure 14. Name the two land uses represented by:
   - the green rectangles
   - the grey areas.

Identifying coastal landforms and land uses

By the end of this section you will:

- be able to identify coastal land forms and land uses from maps and aerial images.

Coasts are used for many purposes. In the past, cliffs made excellent defence sites for castles or larger settlements. Today they have many uses related to tourism, such as sites for car parks, caravan parks, nature reserves, golf courses and walks. Some spits and coastal areas are used for agriculture and of course fishing and trade. The photo and OS map on this spread show a well-known coastal site, the Murlourgh National Nature Reserve.

Tip 

Make a list of all the main things you would expect to see on a map or satellite image to show the main types of land use.

Figure 16 Murlough National Nature Reserve.

Activities

Look carefully at Figure 17 showing Murlough Bay Nature Reserve and the spit and answer the following questions:

1. How wide is the spit at its widest point?
2. Who owns the land on the spit? What is the evidence for your answer?
3. What is the name of the large home on the spit?
4. Which sport could be played in grid square 3832? What is the evidence for your answer?
5. Name the pathway shown on the spit.
6. Which river flows into Dundrum Bay in grid square 4234?
7. What is the direction of the longshore drift along this part of the County Down coastline?
8. Give the six-figure grid reference of the chambered grave.
9. What evidence is there on the map to suggest that visitors may come to the area?
10. Giving evidence from the OS map extract, name three sporting activities young people could engage in when visiting this area.

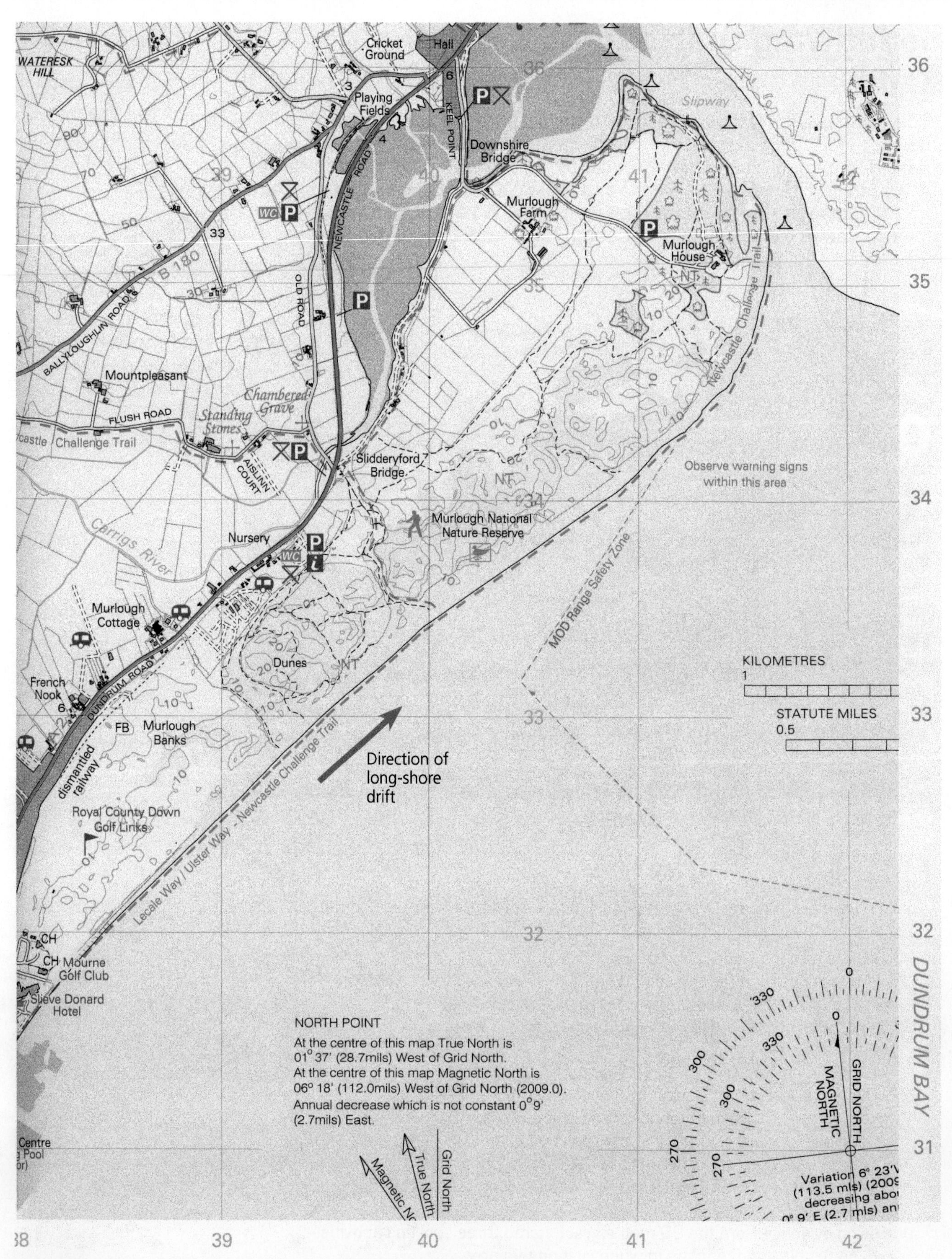

▲ **Figure 17** Murlough National Nature Reserve (Scale 1: 25,000).

Coastal defences

By the end of this section you will be able to:

- explain the need for coastal defences.

Why are coastal defences needed?

The protection of coasts is important as just over half the world's population – around 3.2 billion people – live within 200 km of the sea. By 2025, it is estimated that around 6 billion people will live in coastal areas. Indeed, Africa is the only continent where the majority of people do not live near a coastline.

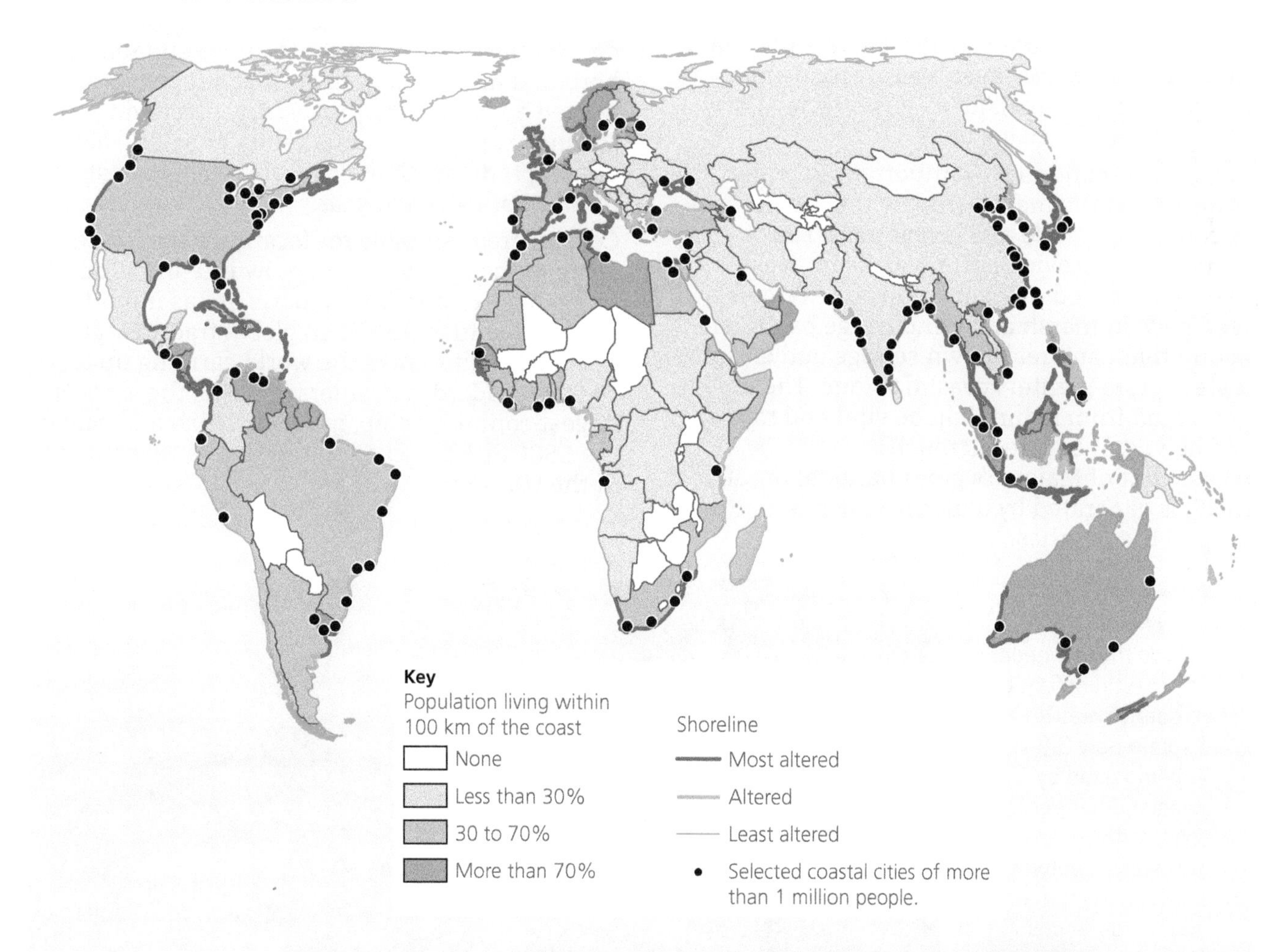

Figure 18 World map showing how most of the world's largest cities have coastal locations and many MEDCs have changed (altered) their coastline from its natural form.

Coastal zones include waters and shores of coastal lands as well as islands, salt marshes, wetlands and beaches. The use of the shoreline has a direct and significant impact on the coastal waters and the large numbers of people that inhabit the areas. Geographical information systems (GIS) are used to understand and control those areas that are likely to be affected by or are vulnerable to rising sea levels or coastal erosion.

Tip

Look at lots of images of the sea defences that you are learning about, such as sea walls. These defences can be many different shapes, so make sure you can recognise them from an image.

Climate change is making the need for coastal defences a pressing issue for many low-lying coastal regions and countries. The average predicted rise in sea levels due to the thermal expansion of water is about 48 cm, although it could be up to almost 90 cm by the end of the century. As sea water becomes warmer the molecules of H_20 move further apart, causing the volume of water to increase (thermal expansion). The World Bank estimates that a 1 m rise in sea levels would flood half of Bangladesh's rice fields and force the migration of millions of people. The European city of Amsterdam is already mostly below sea level, and relies on sea defences to protect it from flooding. The islanders of Tuvalu, in the Pacific Ocean, have already made plans to abandon their homeland as it is now regularly being flooded by rising tides.

Coasts are economically important to many countries as the main ports are the centre of commerce, trade and investment. The marsh areas provide natural areas for waste assimilation and detoxification, for example reed beds in marsh areas encourage bacterial action that can break down sewage and clean the water before it enters coastal waters. The income generated from fishing can be vital and tourism can be very lucrative. As much as 60 per cent of Majorca's gross national product (GNP) is generated by tourism, which is centred around the beaches.

In Dubai, they have gone as far as creating new beach front space by building artificial islands just off the coastline like the Palm Islands in Dubai (see Figure 19).

The fishing industry is still very important to coastal zones, both in LEDCs and MEDCs (Less Economically and More Economically Developed Countries, see page 108). The Food and Agricultural Organisation (FAO) estimates that over 100 million tonnes of fish are caught from the wild each year and assure the livelihoods of 10–12 per cent of the world's population, with more than 90 per cent of those employed by fisheries working in small-scale operations in developing countries. Coastal areas suitable for harbours have fishing fleets and support other related activities like processing, preserving, storing, transporting, marketing or selling fish or fish products. China captures the greatest tonnage of fish each year.

Coastal areas suitable for locating a port were some of the first places to grow into important cities, like London. Today, ports are as important as ever due to increases in global trade. Large container ships cross the world carrying up to 19,000 standard containers. In 2015, the world's largest container ship, made in China and called the *CSCL Globe*, docked in the Port of Felixstowe in the UK.

Activities

1. Why is there a need for coastal defences in some areas?
2. Explain how GIS is being used by coastal engineers.
3. Explain the connection between climate change and coastal defences. Refer to specific places in your answer.

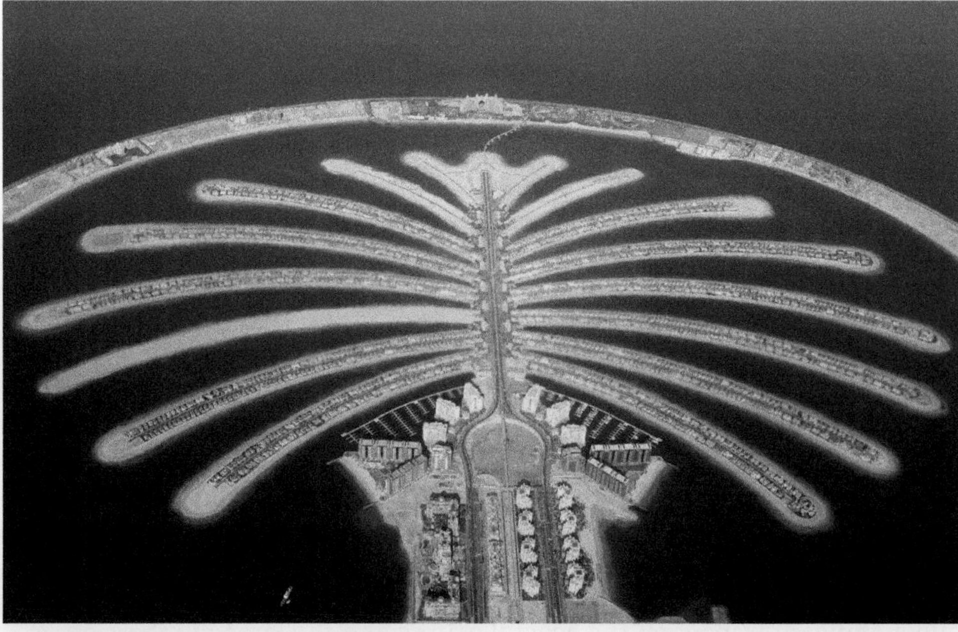

Figure 19 Palm Jumeirah, Dubai.

Methods of coastal management

Coastal areas need management to:

- keep the sea out
- retain cliffs and beaches.

There are two methods of coastal management available to planners: hard and soft engineering methods.

By the end of this section you will be able to:

- describe and evaluate hard engineering methods of coastal management
- describe and evaluate soft engineering methods of coastal management.

What are hard engineering coastal defences?

Hard engineering involves building artificial structures along the coastline to reduce erosion or manage deposition. For centuries, these have been the preferred method of managing British coastlines. They include sea walls, groynes and gabions.

Sea walls

The most common way to keep the sea at bay is to build sea walls. These look like tall concrete walls built at the back of beaches. They may have a curved shape which is designed to deflect the erosive energy of the wave and add extra protection against waves topping the wall.

Sea walls are expensive to build (they cost around £10 million per kilometre), and the need for constant maintenance means costs continue. They can be economically acceptable if they are needed to protect many people and properties, like in Portrush.

Figure 20 The sea wall in Portrush, Co. Antrim.

Activities

1. Figure 21 shows a curved sea wall. Why is it effective in keeping the sea at bay?
2. Why would a sea wall that is more vertical be less effective in keeping the sea at bay?

Figure 21 A curved sea wall in Blackpool.

Groynes

Beaches are essential natural coastal protection and are a main tourist attraction to any coastal area, so many resorts are keen to ensure they are conserved. When longshore drift is displacing sand from a beach, then groynes or beach replenishment are strategies used to ensure beach survival.

Groynes are often made of hard wood and look like low fences stretching seawards out along a beach at intervals of about 50 m. They slow down longshore drift and promote the deposition of sand, thus building up the beach. The wood will eventually weather down, so groynes have a lifespan of around 20 years. They cost around £5,000 per metre to build. Modern construction techniques favour rock groynes, which have a much longer life span than wooden ones.

However, groynes can cause problems too: they can reduce public access along a beach and can cause extra erosion further down the coast as beach material cannot move naturally by longshore drift.

Figure 22 How groynes can change the shape of a beach.

Gabions

Cliffs can be difficult to retain, but recently gabions have been used successfully as a short-term measure to stabilise cliff bases. A gabion is a metal cage, measuring about 1 m by 1 m, that is built on site from six metal mesh sides and then filled with local rocks. As gabions rust and can be damaged during severe storms, they do not provide a long-term solution to coastal management. Such damage is seen in Figure 23, where the nearby gabion baskets on Chilling Cliff in Hampshire have been damaged by a storm. The main advantage of gabions is their low cost.

Figure 23 Collapsed gabions, Hampshire.

What are soft engineering coastal defences?

Soft engineering measures tend to be more sustainable than hard engineering solutions for coastal management. They do not involve the large-scale building required for the hard engineering methods and often take advantage of natural processes to be effective. They have low costs, both economically and environmentally, and so have become the favoured choice for modern coastal engineers.

Beach nourishment

Beach nourishment is sometimes called beach recharging. Sand is dredged from the seabed and added to an eroded beach, or even brought in by lorries to add extra material along a stretch of coastline. It can be cheaper than groynes, costing approximately £3,000 per metre, but it needs more regular maintenance. Nourished beaches erode faster than natural ones because the sand is not as tightly packed. Beach nourishment is very expensive and is only used where there will be very significant economic returns, such as from tourism.

▼ **Figure 24** Beach nourishment: beach restoration efforts on Jennetts Pier, Nag's Head, on the east coast of the USA. The photo on top shows what it looked like in November 2010, and the photo below shows it after nourishment in June 2012.

Managed retreat

Due to economic or environmental costs, in some coastal areas a decision is made to allow nature to take its course, or even to encourage an area of coastline to be changed by coastal processes. This is called managed retreat. Any people living on threatened land may be moved out, farmers compensated for lost land and buildings demolished. One such scheme in the UK was the recent Medmerry realignment coastal scheme. The short-term cost of £28 million was justified as the Environment Agency claims it will significantly reduce the flood risk to communities in the nearby cities of Portsmouth and Southampton.

Activities

1 Complete the following table to show which coastal defence methods are examples of hard engineering and which are soft. One has been completed for you.

Name of defence method	Hard or soft engineering?
Gabion	Hard
Sea wall	
Beach nourishment	
Groynes	
Managed retreat	

2 Choose one hard and one soft coastal engineering strategy. For each one, state its name, describe it and explain how it works.

Weblinks

www.snh.org.uk/publications/on-line/heritagemanagement/erosion/sitemap.shtml – a guide to managing coastal erosion in beach/sand dune systems from the Scottish Natural Heritage.

http://www.bbc.co.uk/news/science-environment-24770379 – a good starting point to find out about the Medmerry realignment scheme.

Weblinks 

www.youtube.com/watch?v=AjZXFw9iWkw – go to this website to see the whole of Newcastle's coastline from the air.

www.newrymournedown.org – the website of Newry, Mourne and Down district council.

CASE STUDY

Coastal management in the British Isles: Newcastle, Co. Down

- Tourism has been the major cause of pressure on the coast at Newcastle, Co. Down. The opening of a rail link between Belfast and the town in 1869 meant that the seaside resort was just an hour's journey from the city. The town is set in a bay, which boasts 8 km of beach that attracts tourists. Boarding houses were built as close to the sea as possible. Even today, Newcastle is a popular day or weekend destination, with thousands packing the beach and main streets on bank holidays. It has been estimated that the population of Newcastle rises by 15,000 in the summer months, and it's all down to tourism.

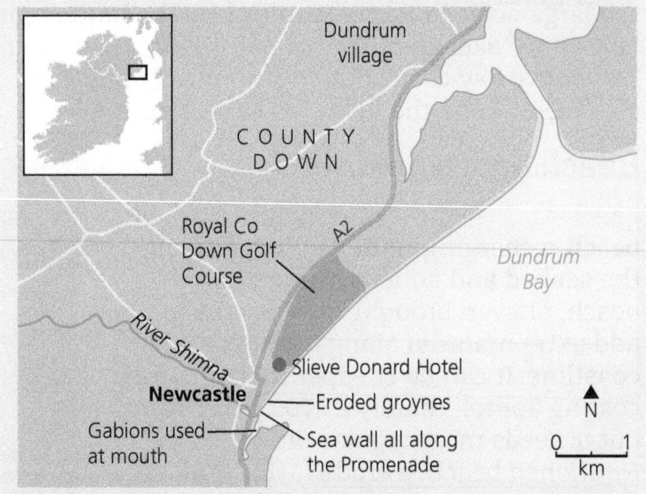

▲ **Figure 26** Newcastle, Co. Down.

Why does Newcastle make a good case study?

- Newcastle is located in Dundrum Bay, which has a 20-km-long beach running between St John's Point and Newcastle Harbour. Waves are the most important driving force controlling the natural coastal system here. These waves are gentle, as there is a limited fetch area, and they approach from the south-east. The shallow and wide beach dissipates wave energy, meaning constructive waves dominate, except during storm times. The sediment which makes up the beach was washed down from the mountains after the last Ice Age. A wave cut notch in Dundrum Village at 14 m above the current sea level is believed to mark the maximum height of the sea during the late glacial period 15,000 years ago. Coastal erosion from the Newcastle coastline provides a sediment source for other spectacular beaches and important ecosystems like the sand dunes of Murlough Bay.

▼ **Figure 25** Gabions and rock armour employed as coastal defences where the Shimna river enters Dundrum Bay, Newcastle, Co. Down.

Gabions (see page 49)

Rock armour (see page 49)

What is the coastal management strategy used at Newcastle?

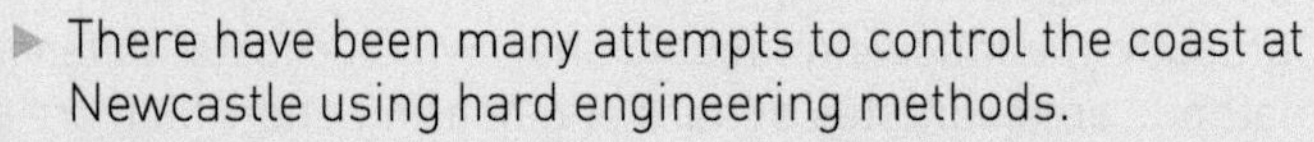

- There have been many attempts to control the coast at Newcastle using hard engineering methods.

Groynes

- In the 1980s, old urban council created concrete groynes near the Newcastle centre section of the beach between the mouth of the Shimna and the Sileve Donard Hotel, to trap and hold sand that was drifting north-east. Since they have now decayed, they can no longer perform this function and may have contributed to sand loss at Newcastle.
- The present council is carrying out a study to see if a new set of wooden groynes could stabilise sand on the beach. Each would be 20–30 m long and would cost £1,250 per metre.

Gabions

- Gabions have been used to protect the recreation ground built over the beach at the mouth of the Shimna, where it enters Dundrum Bay in Newcastle. The first set had badly decayed and were no longer proving effective, so in the regeneration programme in 2006 they were replaced and a new footbridge was built on the stabilised coast to allow unrestricted pedestrian access along the promenade. Gabions are more sustainable than rock armour or a sea wall as they allow water to enter each cage and slowly dissipate the energy rather than deflect it back outwards.

Rock armour

- Rocks may be used to control erosion by armouring a dune face. They dissipate the energy of storm waves and prevent further recession of the backshore if well designed and maintained. Rock armour is used widely in areas with important backshore assets subject to severe and ongoing erosion where it is not cost effective or environmentally acceptable to provide full protection using sea walls. This has been used along various sections of Dundrum Bay. In the late 1990s, extensive rock armouring was constructed to protect the Royal County Down Golf Course. This has proved unsustainable, as it is reducing the sediment supply for Murlough Bay, an Area of Special Scientific Interest (ASSI).

Sea wall

- The need for a sea wall came when Newcastle experienced urban growth and boarding houses were built close to the coast for the sea views. These new buildings at the time needed protection from the high tides and waves. The sea wall constructed in Newcastle is used as support for the promenade.
- The promenade in Newcastle therefore doubles as a sea wall to protect the town. Following a severe storm in 2002, when the old wall was partially washed away, the promenade and wall have been rebuilt and extended at a cost of £4 million. The wall was raised by 1 m from the old Victorian level and now has a curved, wave-return design to stop water splashing over the wall.
- Although this wall protects the built environment, refracted waves appear to be increasing beach erosion. Combined with a lack of material coming from further up the coast, this means that Newcastle is losing beach sand, so it is unsustainable. Indeed, studies show that north of the Shimna river, the high tide beach has been reduced due to promenade construction.

Is the Newcastle coast being managed sustainably?

- Although Newcastle's beach has been badly eroded over the past 50 years, it remains popular with locals and visitors from other parts of Northern Ireland. None of the hard engineering measures is singularly to blame for this erosion, but such inappropriate and un-coordinated development along the coast has increased problems. It is no longer a naturally functioning zone, where erosion and deposition are in balance. In 2006, the Department of the Environment announced an Integrated Coastal Zone Management Strategy for Northern Ireland that applies for 20 years. In this, groups are encouraged to work together towards a more sustainable approach to coastal management. Down Council is considering building new groynes and even beach nourishment in order to maintain a wide and sandy stretch of beach at Newcastle to satisfy the demands of tourists.
  In 2012, coastal expert Professor Derek Jackson from Ulster University noted that sea walls like the one built in Newcastle simply deflect the waves and gradually erode away once stable beaches.
  He also commented on the lack of sustainability of the current sea defences all over Northern Ireland, especially given a background of sea level rises related to climate change.

Figure 27 Newcastle as it is today.

Figure 28 This photograph of Newcastle harbour was taken in 1880.

Sample examination questions

Tip

Make sure you tackle both the 'describe' and 'evaluate' command words here. When evaluating you should be using terms such as 'overall' or 'on balance' in your concluding statement. Remember to include good case study detail (which might include facts and figures) in your answer.

1 Complete the following sentences by underlining the correct word. [4]

(i) Waves are caused by **wind / headlands** shaping the surface of the water.

(ii) Destructive waves have a stronger **swash / backwash**.

(iii) Wave-cut platforms are found at the **base / top** of a cliff.

(iv) Sandy beaches are **flatter / steeper** than those made from shingle.

2 Name two coastal features created by deposition. [2]

3 Choose two of the coastal erosion processes from the box below and explain how they operate. [4]

attrition	corrasion	hydraulic action	corrosion

4 Complete Figure 1 to show how longshore drift would operate on that beach. [3]

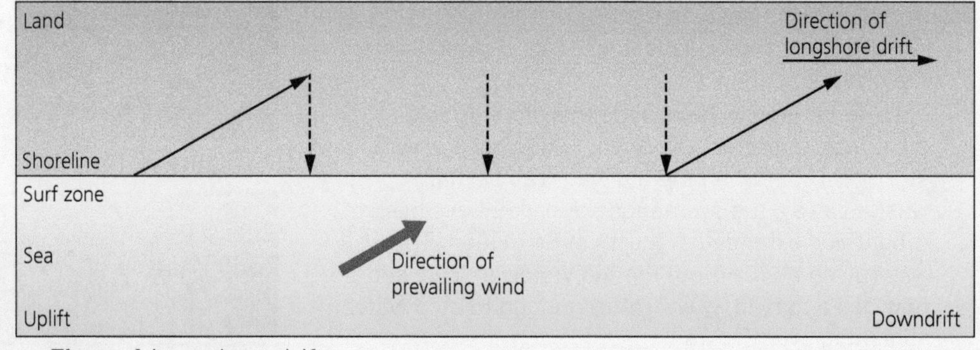

Figure 1 Longshore drift.

5 Study Figure 2. State which colour of arrow shows where coastal headlands will form. [1]

Figure 2 Before and during erosion of the Dorset coastline, UK.

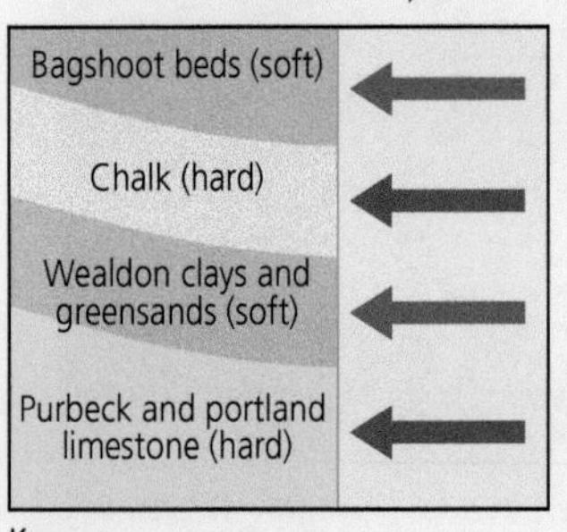

6 Explain the formation of stacks, such as those shown in Figure 3. [6]

Figure 3 A row of stacks.

7 Using Figure 4 on page 52:

(i) State the direction of longshore drift at Murlough. [1]

(ii) State two land uses that can be found on the spit at Murlough Bay. [2]

(iii) Give the distance between the car parks at GR394338 and GR411353. [2]

(iv) Suggest a reason to explain why a developer was not allowed to build a large number of holiday homes in grid square 4033. [3]

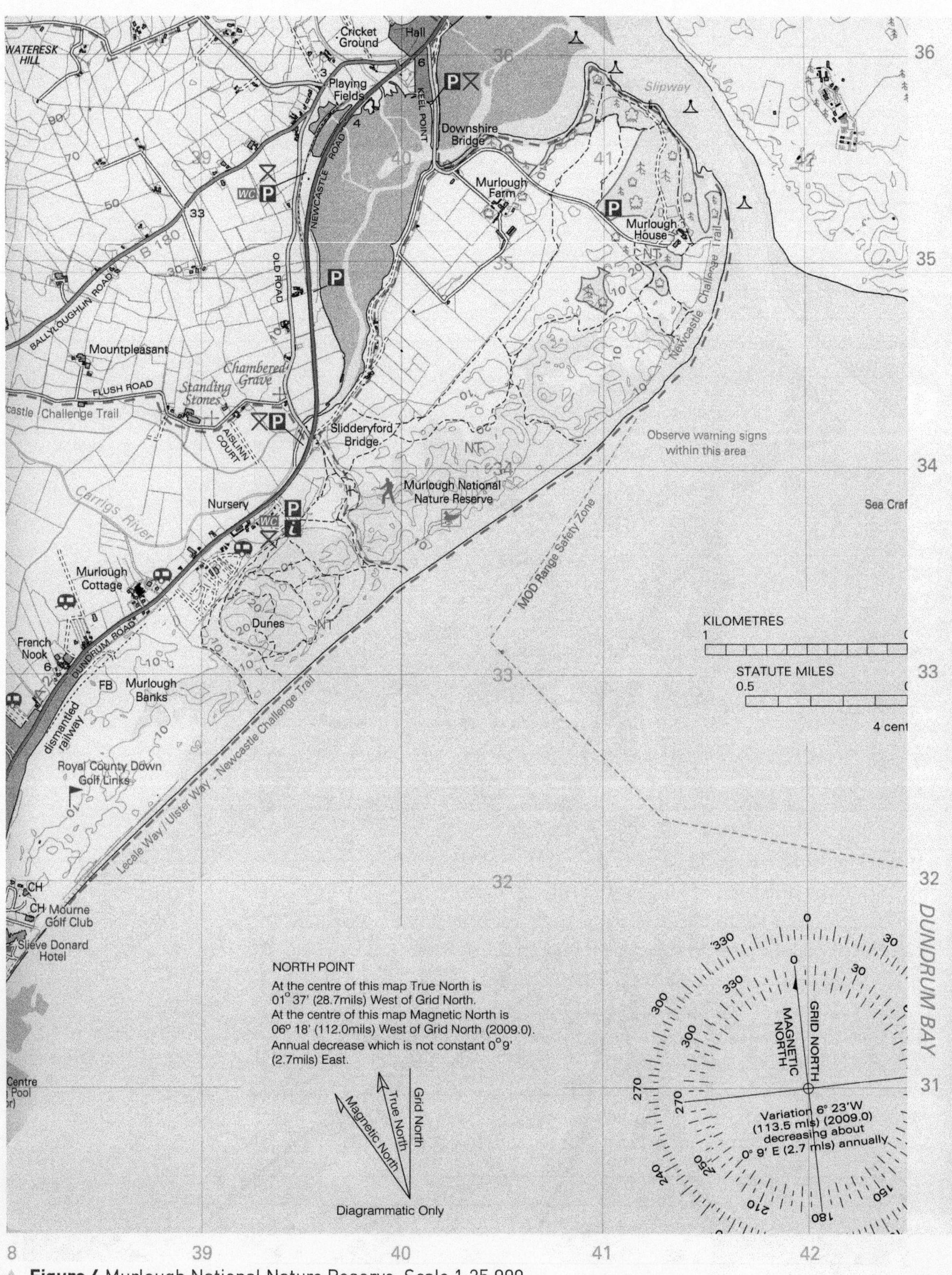

Figure 4 Murlough National Nature Reserve. Scale 1:25,000.

(v) Explain the formation of a spit, such as the one shown here at Murlough. [5]

____

____

____

____

____

8 State three reasons why a place might need coastal defences. [3]

____

____

____

____

____

____

9 Complete the table below to show which methods of coastal management involve hard or soft engineering. One has been completed for you. [3]

Hard engineering	Methods of coastal management	Soft engineering
←	Sea wall	
	Beach nourishment	
	Gabions	
	Groynes	

10 Choose one of the methods in the table above and explain how it helps to defend the coastline. [4]

____

____

____

____

____

____

____

____

11 State the meaning of the coastal management method of **managed retreat**. [2]

____

____

____

____

12 State one advantage and one disadvantage of managed retreat as a method of coastal management. [4]

13 Explain why coasts are described as dynamic environments. [4]

14 With reference to a case study of coastal management in the British Isles, describe the coastal defences used and evaluate their sustainability. [9]

UNIT 1

UNDERSTANDING OUR NATURAL WORLD

THEME C: Our Changing Weather and Climate

Satellite Image of Storm Doris.

Why do weather systems like Storm Doris bring wet and windy weather?

1 Measuring the elements of the weather

By the end of this section you will:

- know the difference between weather and climate
- be able to describe how to measure the main elements of the weather.

Tip

Reduce some of the key ideas discussed here into golden rules, such as: winds blow from high to low.

Weather and climate

What is the difference between weather and climate?

The word weather is used to describe the day-to-day changes in the conditions of the atmosphere – that is, the elements of the weather. Climate is the average conditions of the weather taken over a long period of time, usually 35 years. The climate of Northern Ireland, for example, is described as mild and damp with few moments of extremely low or high temperatures or large amounts of rain. Weather conditions change every day and can vary over short distances, even within Northern Ireland. For example, Belfast can have very different weather to Enniskillen or Coleraine.

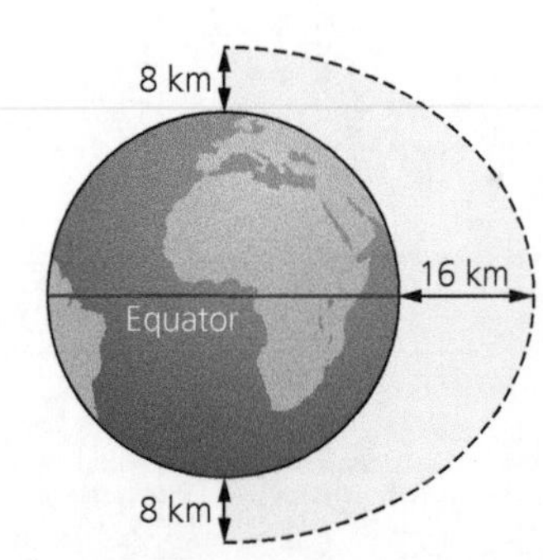

Figure 1 Weather takes place in the lowest layer of the atmosphere.

How do we measure the elements of the weather?

The following table shows some weather elements, their unit of measurement and the instrument used to record the weather.

Element	Unit	Instrument
Temperature	degrees centigrade (°C)	Digital thermometer
Precipitation	millimetres (mm)	Rain gauge
Wind speed	knots or kilometres per hour (kph or km/h)	Anemometer
Wind direction	8 compass points	Wind vane
Pressure	millibars (mb)	Barometer

Figure 2 Ways to record weather.

Digital maximum-minimum thermometer

Barograph

A Stevenson screen stores instruments

Barometer

Rain gauge

Wind vane

Anemometer

Why does a weather station look like it does?

Meteorologists are scientists who study and predict weather. They rely on the data provided by weather stations all over the world to help them. The following information is about one source of these data: land-based weather stations.

When deciding where to locate weather instruments, there are certain things the meteorologists take into account to ensure the accuracy of their data.

Temperature

Temperature is how hot or cold something is. It is measured in degrees Celsius (centigrade) in the UK, although in the USA you will also see temperature in degrees Fahrenheit. When trying to measure weather, it is important that we know the air temperature.

Traditional thermometers contained mercury, which expands with rising temperatures. More recently, mercury was replaced by alcohol with red or blue dye in standard glass thermometers. However, today most households and businesses have digital thermometers as they are safer, easier to read, and work faster. Digital thermometers contain an electric resistor, also known as a thermistor, which is temperature-sensitive. When the temperature rises, the thermistor becomes more conductive. A microcomputer pinpoints the temperature by measuring the conductivity, and displays it on an LCD screen which the user can simply view as a reading. Often such thermometers can be programmed to record the highest and lowest temperatures over a 24 hour period and can display these separately.

As temperature varies depending on the type of surface or exposure to sunlight, meteorologists have agreed to standardise the measuring of temperature to allow for comparison. The shade temperature is the agreed correct measure. In a weather station, thermometers are placed within a Stevenson screen. This is set on stilts above the ground to ensure that it is air and not ground temperature that is recorded. The box shades the instruments from direct sunlight and the slatted sides allow the free flow of air. The Stevenson screen is painted white to reflect sunlight and is located on open ground away from buildings.

Today we are used to knowing the temperature because we have it displayed in cars, on smartphones, on central heating controls and even on the side of buildings. These sources use digital thermometers. It can be easier to read a digital display as small changes in tenths of a degree of heat might be missed using a traditional thermometer.

Figure 3 This thermometer shows the kind of circuit used in most digital thermometers, it includes resistors, a digital display and thermistors. The resistance of thermistors is dependent on temperature. This digital thermometer is displaying air temperature in degrees Fahrenheit. It is alternative measure of heat to Degrees Celsius and is often used in the USA. 1°F is about -17°C.

Precipitation

The term precipitation includes all types of moisture in the atmosphere, from rain to snow and hail to fog.

A rain gauge is a cylinder that catches precipitation and funnels it into a measuring flask. It is located in an open area to avoid shelter that trees might provide. Being sunk into the ground avoids excessive evaporation and provides stability in windy conditions. Each day the flask is taken out and the amount of precipitation is recorded in millimetres (mm) by transferring the liquid into a small measuring cylinder.

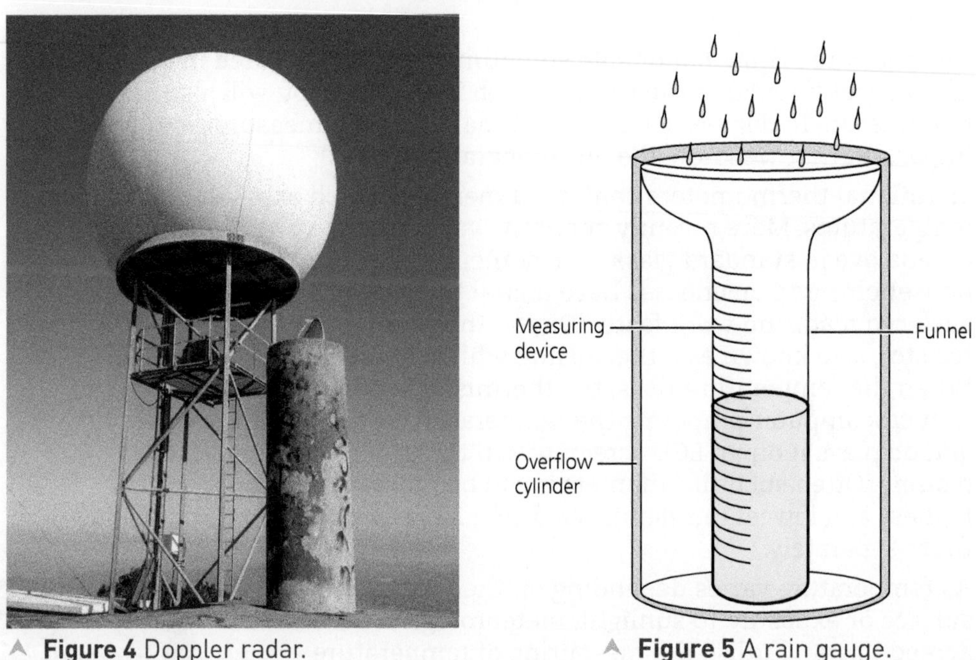

Figure 4 Doppler radar.

Figure 5 A rain gauge.

Modern weather forecasts on television or streamed through internet sites often show rainfall as a shadow superimposed onto a map. This information was probably originally obtained from a rainfall radar. The Met Office tracks current rainfall and produces real-time digital rainfall maps of the UK.

Wind

Air moves around in the atmosphere, which we feel as wind. It is caused by differences in air pressure across the planet's surface. Winds are movements of air from high to low pressure.

Anemometers have three cups mounted on a high pole to catch the wind. As the wind blows, the cups spin and the wind speed is recorded on a dial which can be read a little like the speedometer in a car. Wind speed is measured in knots or in km/h.

Wind direction is shown by a wind vane. The top section is loose and moves with the wind, while the base is fixed and orientated to show the main eight points of the compass. The arrow points to the direction from which the wind is blowing.

The Beaufort Scale is used mostly in marine weather forecasts given by the Met Office. It describes wind intensity based on observed sea conditions, or on land by the effect on rising smoke and tall standing trees. Beaufort Scale 8 (Gale Force) is sometimes referred to in other forecasts, as at this level the wind is strong enough to cause damage to property.

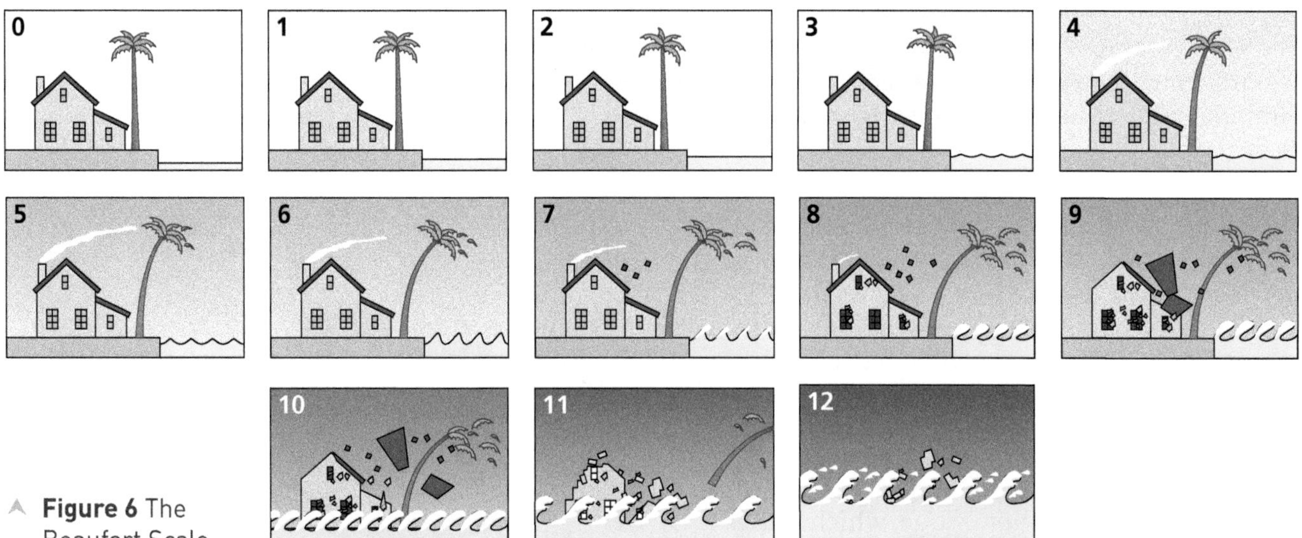

Figure 6 The Beaufort Scale.

Atmospheric pressure

Pressure is the weight of a column of air. The average pressure at sea level is 1012 mb. The barometer is used to measure atmospheric pressure, in millibars. There are two main types of barometers: the mercury barometer and the aneroid barometer.

The mercury barometer is composed of a glass tube about 1 m tall with one end open and the other end sealed. The tube is filled with mercury. This glass tube sits upside down in a container, called a reservoir, which also contains mercury. The mercury level in the glass tube falls, creating a vacuum at the top. The barometer works by balancing the weight of mercury in the glass tube against the atmospheric pressure. If the weight of mercury is less than the atmospheric pressure, the mercury level in the glass tube rises. If the weight of mercury is more than the atmospheric pressure, the mercury level falls.

Aneroid barometers are commonly found in homes. The aneroid barometer is operated by a metal cell containing only a very small amount of air. Increased air pressure causes the sides of the cell to come closer together. One side is fixed to the base of the instrument while the other is connected by means of a system of levers and pulleys to a rotating pointer that moves over a scale on the face of the instrument. This pointer is usually black and is used to display the atmospheric pressure.

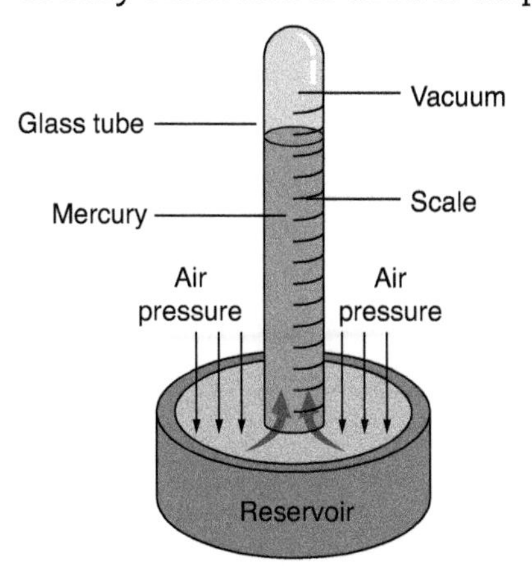

Figure 7 A mercury barometer.

Figure 8 An aneroid barometer.

Activities 

The Beaufort scale (see Figure 6), named after the Northern Ireland man who invented it in 1806, is often used to describe wind speed. Find out the answers to the following questions:

1. How many levels are there to the Beaufort scale and what are they called?
2. What wind speed (in mph or km/h) defines each level of the scale?
3. What indicators, such as the movement of smoke or branches, are associated with each force on the scale?
4. Draw a table to record your answers.

Clouds

Cloud watching can be fun, especially in Northern Ireland, where we get a wide variety of cloud types.

There are three basic types of clouds:

- Cirrus clouds. These are the whitest, highest clouds made of tiny ice crystals. They are often wispy in appearance.
- Cumulus clouds. They are often low in the air and look like cotton wool or like cauliflower on top with a flat base. Cumulus is the Latin word for 'heap'. Clusters of small, white cumulus clouds are usually a sign of fine weather. Sometimes, cumulus clouds develop into the storm cloud cumulonimbus, which brings lightning and thunder. Cumulonimbus clouds are called 'the king of clouds'. The base of a cumulonimbus cloud is often low but it may be as high as 10 km.
- Stratus clouds. These appear as light grey clouds that look like even sheets and cover all or part of the sky. They are composed of fine water droplets that become larger as they collide with each other and are often very low in the air.

Nimbus refers to any cloud that is rain-bearing. You can tell this by the grey colour of the cloud. You can identify clouds by looking at their shape and height above the Earth. Cloud cover is measured in oktas, that is, eighths of the sky covered with cloud.

Figure 9 Fair weather cumulus clouds.

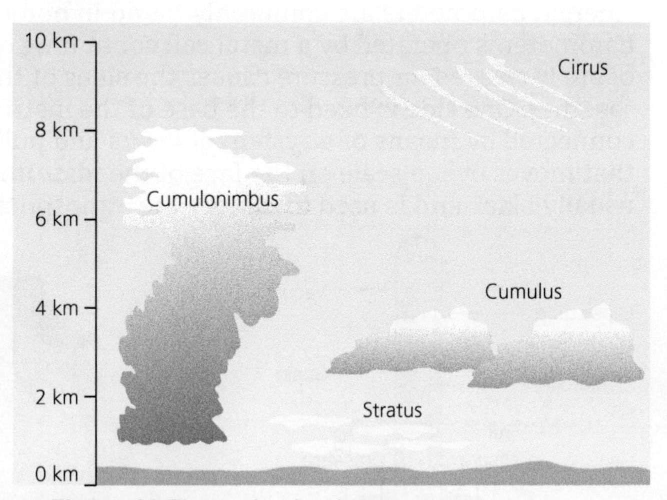

Figure 10 The main cloud types.

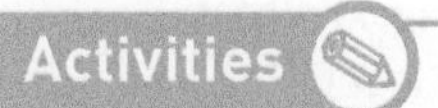

Activities

1. Explain the difference between weather and climate.
2. Make a copy of the table showing how weather elements are measured on page 56, missing out a label in each row. In class, try to finish your partner's table with the correct labels.
3. Explain why Stevenson screens look the way they do.
4. Describe how a rain gauge works.
5. Describe how a wind anemometer works.
6. What indicators, like the movement of smoke or trees, are associated with levels 0, 3, 10 and 12 on the Beaufort Scale?
7. Create a true/false set of statements about the main cloud types. Swap with a partner and see if you can answer all their questions correctly.

Sources of data for weather forecasting

A weather forecast is made using computers and the records of past weather patterns to predict current weather. Forecasters use data collected from many sources to produce a synoptic chart, which shows the predicted weather conditions.

Weather forecasts are usually accurate for a period of 24 hours and reasonably accurate for up to five days ahead. Beyond this, they become increasingly unreliable. The chaotic nature of our atmosphere means that it is unlikely that we will ever be able to make accurate long-range weather forecasts.

By the end of this section you will be able to:

- describe the sources of data used to produce weather forecasts.

Tip 

Be prepared to describe differences between the sources of weather data by writing a list of these differences.

Skills activities 

Work in groups to complete the following fieldwork activities.

1. Take weather readings every day for a month using the instruments shown in Figure 2.
2. Devise a suitable data collection sheet.
3. Record your measurements and observations.
4. Using ICT, produce appropriate graphs to represent the data you have collected (temperature, precipitation, pressure, wind speed, wind direction, cloud cover, cloud type).
5. Look closely at your table of data and observations and the graphs you have produced and answer the following questions:
   - Can you identify any relationships between weather variables, e.g. pressure and temperature?
   - If so, what are the relationships you have discovered about the climate around your school?
   - Did any of your findings surprise you?
   - If your school was to set up all the instruments to take weather readings, where would you locate them and why?
   - Do you think there are any microclimates around your school? Did you find any evidence to support your view?

How do land-based stations help forecast the weather?

Weather forecasting began in the late 1700s using only a few unreliable instruments. Today there are more than 10,000 land-based weather stations located in countries all over the world. They collect data on the elements of weather every synoptic hour (every three hours) every day of the year. This can be done manually or by using a digital display.

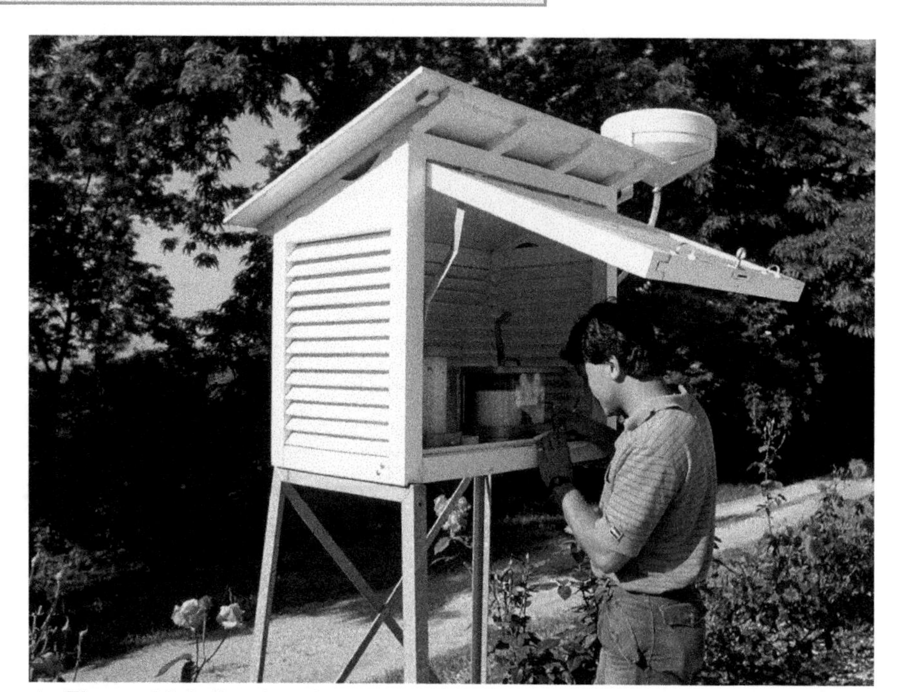

Figure 11 Collecting data.

How do sources of data in the air help forecast the weather?

Most weather data is captured in digital form from space, using satellites. The data is then used by organisations such as the National Weather Service who release it in GIS (geographic information system) format to other local weather forecasting services. Such data goes beyond simple temperature recordings, it also covers snowfall, wildfire locations and Doppler radar images of precipitation.

Satellites are small spacecraft that carry specific weather instruments. They are launched into space and orbit the Earth recording its weather data. The radiometer provides colour images of clouds. The scatterometer uses microwaves to detect the speed and direction of winds.

If a satellite is positioned over one place and moves at the same speed as the Earth, it is called a geostationary satellite. It only provides data for that one place. Polar satellites offer the advantage of daily global coverage, by making nearly polar orbits roughly 14.1 times daily. Currently in orbit we have morning and afternoon satellites, which provide worldwide coverage four times daily. These satellites are able to collect global data from different places on a daily basis. The ocean temperature information they are gathering is an important source of evidence to document climate change.

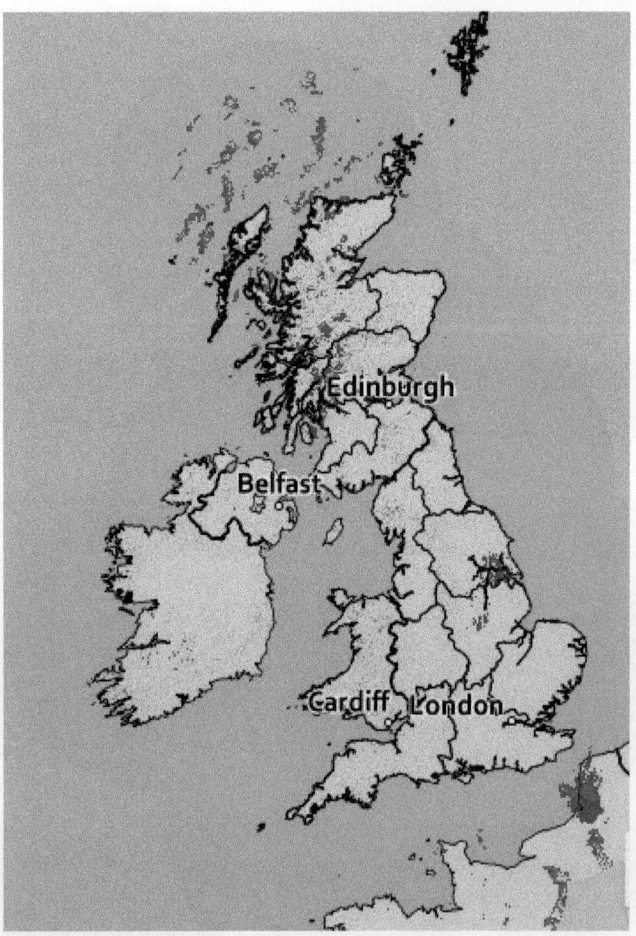

Figure 12 Rainfall radar image showing showers over North West Scotland and some rain over Kingston upon Hull in Eastern England.

Drones and rainfall radar

Drones are a new and evolving technology. Meteorologists see potential in using them as a source of weather data that will fill the gap between the surface, land-based systems and satellites. They can be controlled, unlike weather balloons, and can be flown into hurricanes and tornadoes to collect information on all the elements of weather. Large drones are called Unmanned Aerial Vehicles (UAVs). Experts believe that the use of drones will be especially useful at helping to improve warning times for extreme weather events, like tornadoes, from the current 20 minutes up to 60 minutes. Current weather forecasts are becoming more precise, not only with regards to general weather, but on when and how much rainfall an area might expect.

Rainfall radar maps are available on the Meteorological Office website to show current and forecast rainfall patterns using a colour coded key for rainfall amount. Radar units are made up of a transmitter and a receiver. The transmitter sends out harmless microwaves in a circular pattern. Rainfall scatters the waves, sending some back to be picked up by the receiver. The pattern of scattering can be mapped to create a rainfall image for a wide area such as the UK.

Figure 13 A weather satellite orbiting the Earth.

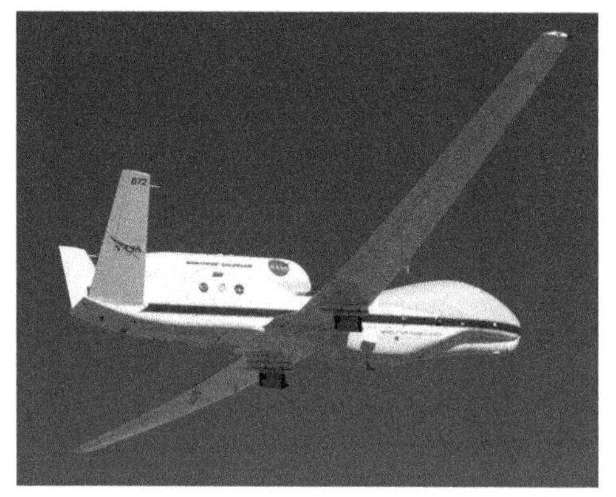

Figure 14 Weather researchers hope drones like NASA's Global Hawks can take hurricane tracking to a new level.

How do sources of data in the sea help forecast the weather?

The main sources of weather data at sea are weather buoys stationed in mid-ocean locations. They transmit weather data via satellite to weather centres for use in forecasting and climate study. Both moored buoys and drifting buoys (drifting in the open ocean currents) are used.

Additionally, most ships have a weather station attached to them. These weather ships record and send data in the same way as the buoys. Mariners need to know future weather hazards.

Figure 15 A weather buoy.

Figure 16 Weather instruments on a ship.

Activities

1. Explain the difference between a geostationary and a polar satellite.
2. Create a mind map to help illustrate the main sources of weather data collection from the air.
3. Suggest why sailors need to have accurate weather forecasts.

2 Factors affecting climate

By the end of this section you will be able to:

- describe and explain the main factors that affect the climate of an area.

Tip

Try using an mnemonic like PLADS, which stands for Prevailing wind, Latitude, Altitude and Distance from the Sea, to help you memorise these factors.

Types of climate

There are different types of climate, ranging from dry desert climates to the four-season temperate climate which we experience in the UK. Climate only considers temperature and rainfall of an area, shown as a climate graph.

Which factors affect climate?

There are four main factors that affect climate: latitude, prevailing wind direction, distance from the sea and altitude.

- **Prevailing winds** will bring different amounts of moisture and heat depending on where they have come from. Places that have prevailing winds from dry, hot continental interiors tend to end up with a desert climate, like the Atacama region of South America. For the UK, our prevailing wind direction is south-westerly, so we have a mild but damp climate.
- **Latitude** affects the relative position of the Sun in the sky. Places near or at the equator are warmer than places near the poles because they get stronger, more direct radiation from the Sun. Also, the radiation travels through less atmosphere, and so there is a reduced chance that it will be reflected back by clouds or other particles in the air.
- If you have ever seen a snow-capped mountain the impact of **altitude** on climate has been clear. Upland areas tend to be wetter than lowland places as the air is forced over the mountains, cools and water vapour condenses to create clouds and rain. Also, temperatures fall as height above sea level increases, by about 1°C per 100 m.
- **Distance from the sea** (continentality) influences temperature range. Areas close to the sea tend to have a smaller annual temperature range than areas inland. This is because the sea (as a liquid) heats up less quickly than the land, but holds the heat it gains for longer. In the winter months, the warmed sea surrounding coastal areas spreads heat to the nearby land, meaning they are warmer than inland areas.

Activities

1. Name the prevailing wind direction for the UK and describe the usual weather it brings.
2. If it is 16°C in Newcastle and I want to climb Slieve Donard, which is 850 m high, what should I expect the temperature at the summit to be?
   Tip – remember that temperature falls about 1°C per 100 m.
3. Explain why it is always colder at the poles than the equator.

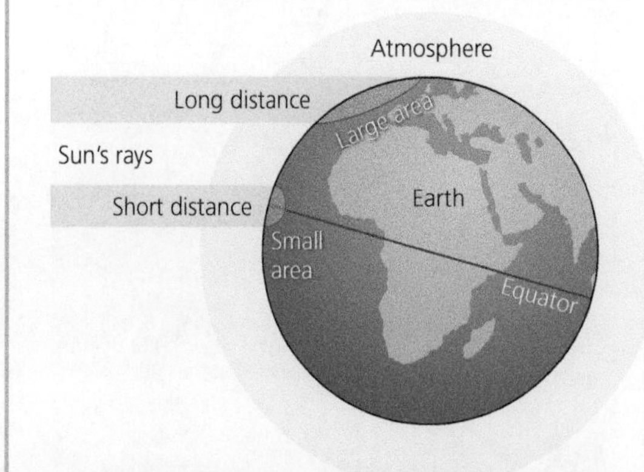

Figure 17 How latitude affects climate.

Figure 18 Glaciers and snow exist at the equator only on the highest peaks, like Mount Kilimanjaro. Here the effects of altitude outweigh latitude.

Air masses affecting the British Isles

What are air masses?

The British Isles are affected by four main air masses, and this makes the weather very changeable. An air mass is a large body of air with similar temperature and moisture characteristics all the way through it.

The characteristics of the air mass depend on where it comes from or the source region.

Figure 19 Characteristics of air masses.

Air mass	Direction	Moisture characteristics	Temperature and seasonal variation
Tropical maritime	South-westerly	Picks up moisture from the sea surface and so brings wet weather to the British Isles	Mild and wet in winter, warm and rainy in summer
Tropical continental	South-easterly	Dry because it forms over land surfaces	Mild and dry in winter, hot and dry in summer. More common in summer than winter; can create heatwave conditions
Polar maritime	North-westerly	Wet, because it picks up moisture as it travels over the Arctic or North Atlantic Ocean	Cold and wet in winter, cool and damp in summer. Rarely experienced outside the winter months
Polar continental	North-easterly	Dry, as it comes over Northern Asia, a landmass	Cold and dry in winter, hot and dry in summer. Mainly affects the British Isles during the winter half of the year.

Where air masses meet, a front is formed. A front separates warm and cold air masses. The tropical maritime air mass is usually found between the warm fronts and cold fronts (in the warm sector) of depressions.

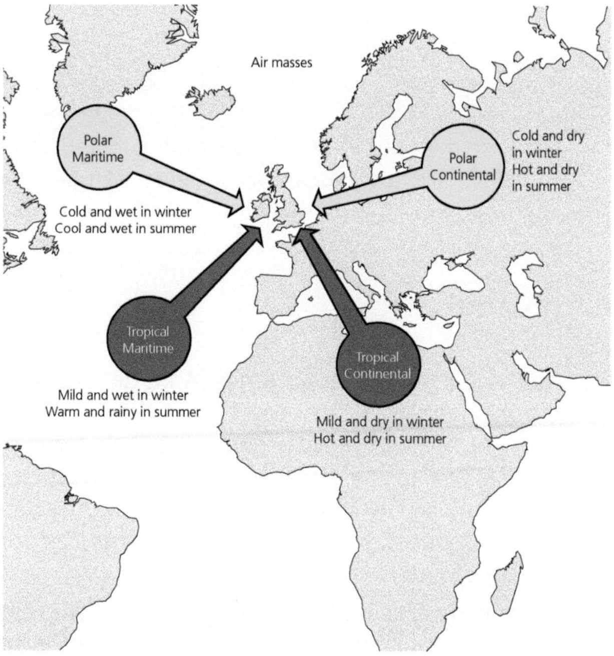

Figure 20 Map of the British Isles showing the directions of the main air masses.

By the end of this section you will:

- know and understand the temperature, moisture characteristics and seasonal variation of the four main air masses that affect the weather of the British Isles.

Tip 

Learn the characteristics of the four air masses carefully and be prepared to compare and contrast them in a question.

Activities 

1. What is an air mass?
2. Name the two dry air masses that affect the British Isles.
3. Describe and explain the temperature of tropical air masses.
4. From what you now know about air masses, suggest why the saying 'When the North Wind doth blow, there shall be snow' might have come about.

By the end of this section you will be able to:

- describe and explain the sequence of weather associated with the passage of a depression
- describe and explain the weather patterns associated with anticyclones during winter and summer.

Tip 

Isobars are lines on a weather map that join places of the same air pressure together.

Tip 

Learn the sequence of weather in a depression carefully, remembering to read the cross-section diagram backwards as places first experience a warm front, then a warm sector, then a cold front as a depression moves over them.

Weather patterns

There are two main types of weather system that cause our weather patterns: depressions and anticyclones.

What are the weather patterns associated with a frontal depression?

Depressions are systems of low pressure; they are like whirlpools of air that develop in the main stream of air movement that comes towards the British Isles from the west. They are areas of low pressure, which generally move towards the east. The winds blow anticlockwise and into the centre of the low pressure. The air rises in the centre.

On weather maps (synoptic charts) a depression has a circular pattern of isobars, with the lowest pressure in the centre and the winds blowing into the centre. Depressions can be hundreds of kilometres wide. Depressions have fronts (hence the term frontal depression) because they form when a warm tropical air mass meets a cold polar air mass. A front divides the two air masses. The warm front and the cold front are separated by a wedge of warm tropical air called the warm sector. All around the warm sector is the cold sector which consists of polar maritime air.

Research activities 

The idea that northerly winds (winds from the north) are cold, and southerly winds (those from the south) are warm (at least in the northern hemisphere) is quite common. Similarly, air that has travelled over the sea picks up moisture, while air travelling over the land is relatively dry. These simple concepts help in the understanding of air masses. However, as may be expected, there are variations on this theme.

1. Find the surface pressure chart for Europe at www.metoffice.gov.uk
2. Work in groups to complete the following activities:
   - On a blank map of Europe mark the main areas of high and low pressure.
   - Draw on arrows to show air flowing from areas of high pressure to areas of low pressure.
   - Look up temperatures of some places that seem to have northerly and southerly winds.
   - Use the data to see if the original idea (hypothesis) was correct, that northerly air masses bring colder temperatures.

Figure 21 Depressions bring wet and windy weather.

The sequence of a depression from an observer's point of view

- *As the warm front approaches*. As a depression approaches, a person at ground level will first see cirrus cloud, high up in the sky. There is no rain yet but temperatures are cool and winds may be strong and from the east or south. As the warm front approaches, the cloud thickens and close to the warm front there will be rain and drizzle as warm air is being forced to rise. Pressure is high but decreases towards the centre of the low pressure.
- *In the warm sector*. Here, temperatures increase in the warm tropical air. There is low stratus cloud but it is mainly dry; this is because water

vapour can easily be held in the warm tropical air without condensation taking place. Wind direction becomes more south-westerly and wind speed usually increases. Pressure values are lowest in this central part of the depression.

- *At the cold front.* As the cold front passes, temperatures fall and the winds will change direction and blow from the north-west. The observer will see towering high cumulonimbus clouds at the cold front and there will be heavy rain. This is because the warm air is rising quickly at the steeply sloping cold front. Pressure starts to rise after the cold front as the depression passes. As the cold front moves away there will be scattered showers from some isolated cumulus clouds; the wind speeds become lighter.

A depression ends when all the warm tropical air in the warm sector is lifted off the ground. This is shown on a weather map by an occluded front. This can happen when the cold front moves eastwards faster than the warm front and so cold air pushes the warm front off the ground. An occluded front brings similar weather to a warm front.

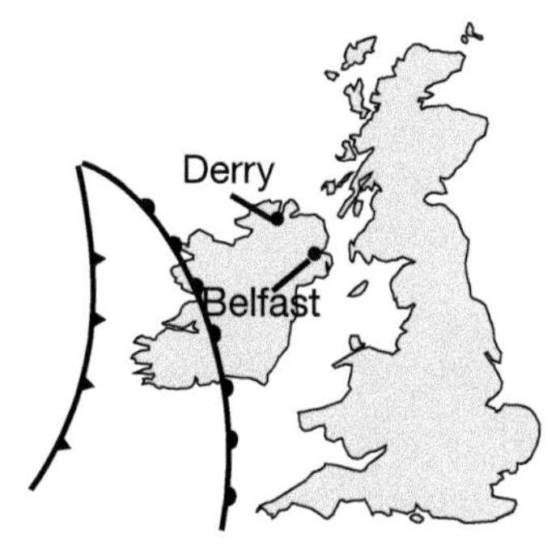

1. A warm front approaches. Skies change from clear, to having high wispy clouds, to developing thicker, lower clouds. Eventually it starts to drizzle.

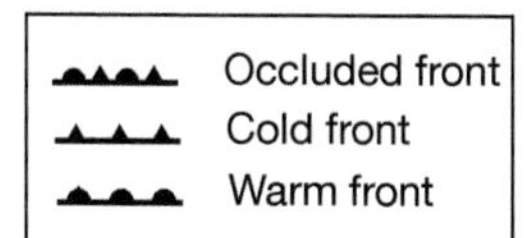

2. The warm front passes over. Drizzle is replaced by steady rain.
The temperature rises in the warm sector. The sky remains grey and overcast. Rain or drizzle continues. The wind blows from the south-west.

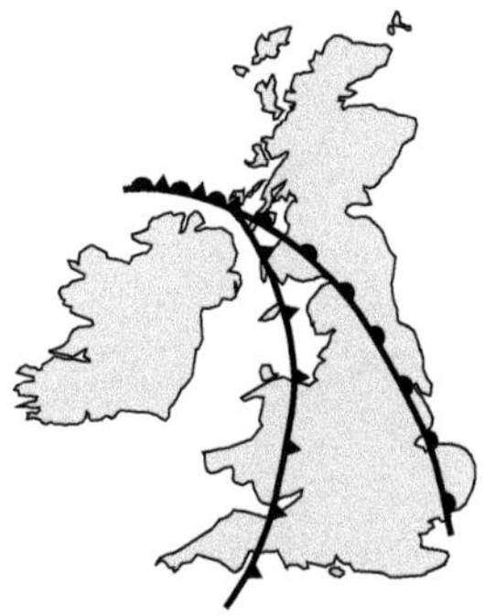

3. The cold front has passed over Northern Ireland resulting in heavy rain and gusty conditions. Temperatures drop as the front passes over. The wind swings around to the north-west. Behind the cold front, the sky clears. The weather now brings sunny intervals with heavy showers.

Figure 22 The sequence of weather as a depression passes.

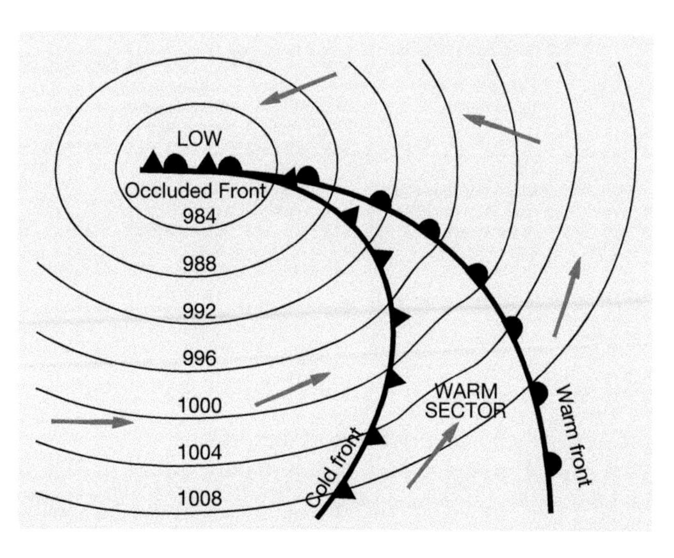

Figure 23 Plan view of a depression. The isobars bend abruptly at the fronts; the winds blow anticlockwise and into the centre of low pressure. The air in the warm sector is rising.

Activities

Study pages 65 and 66 and then answer the following questions. Your answers should be illustrated with simple sketches/ diagrams.

1. What is a front?
2. What is a warm front and how is it shown on a weather map?
3. What is a cold front and how is it shown on a weather map?
4. What is an occluded front and how is it shown on a weather map?

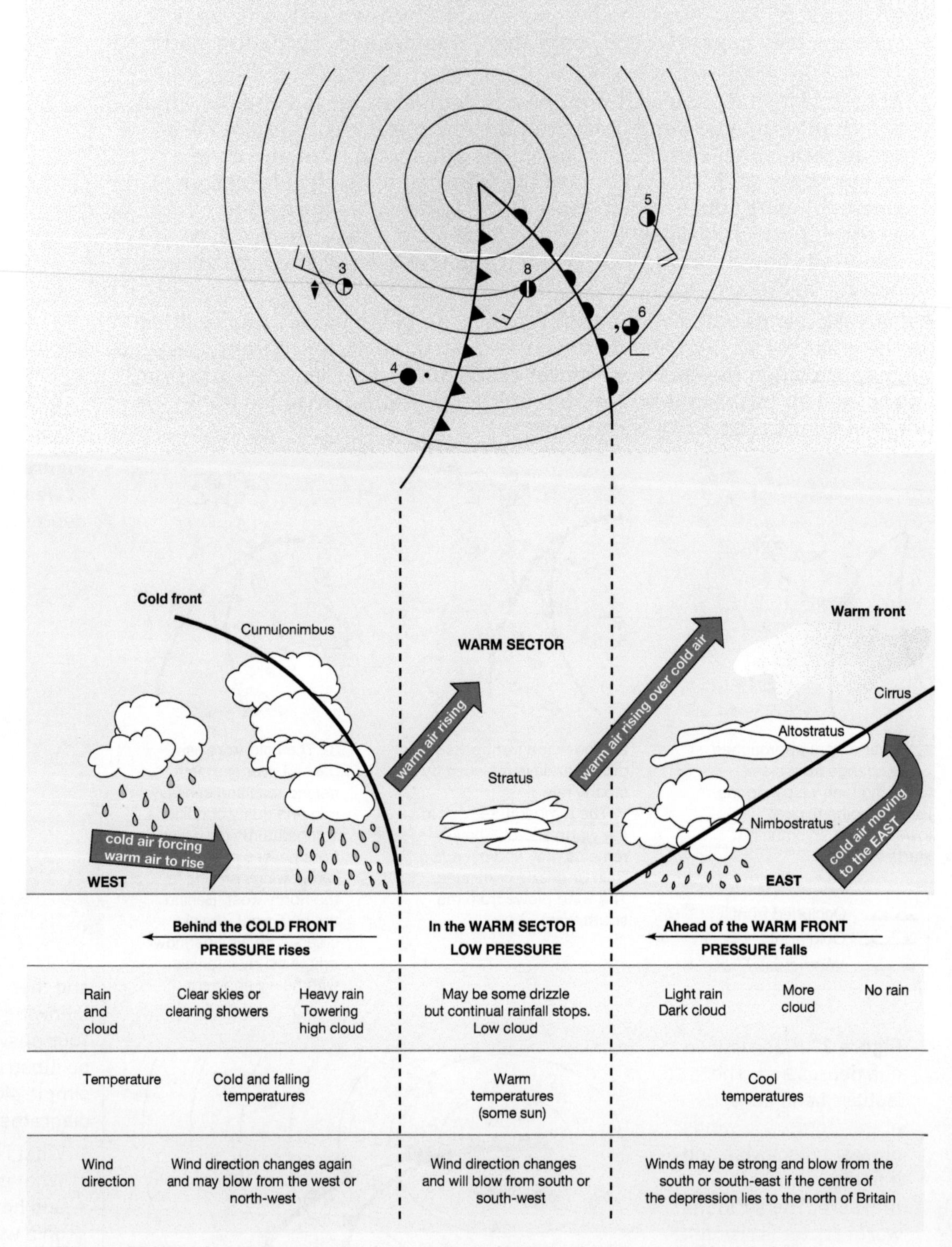

Figure 24 How the plan view of a depression relates to the cross-sectional view.

Activities

1. List three general characteristics of a depression.
2. Describe the weather conditions inside the warm sector of a depression.
3. Describe how and explain why temperature changes as a cold front passes over an area.

What are the weather patterns associated with anticyclones?

Anticyclones are systems of high pressure. In the centre the air is sinking slowly from great heights; as the air sinks, it swirls in a clockwise direction and spreads out at the surface. The sinking air is compressed and warms up as it nears the ground; this means the air can hold more water vapour without condensation taking place and so clouds do not form and it is less likely to rain. This means anticyclones are associated with dry, bright weather.

In an anticyclone, the isobars are spaced well apart and so the pressure gradient is gentle; this means the wind speeds are low and there may even be calm conditions with no wind in the centre of the high pressure. An anticyclone has no fronts and moves very slowly so the weather conditions may not change very much as this system passes across the British Isles.

A summer anticyclone brings cloudless skies and bright sunshine and high temperatures during the day. At night, temperatures can fall due to rapid cooling caused by heat escaping through radiation into the atmosphere when there are no clouds. The mist created is usually easily evaporated by the strong sunshine in summer.

A winter anticyclone often brings fog or mist; these form at night when rapid cooling occurs and heat is lost by radiation due to the lack of cloud cover. Water vapour in the cold layer of air near the ground condenses and the water droplets are suspended in the air as mist or fog. In winter, the low angle of the Sun means that the rays cannot disperse the fog or mist and it may persist all day.

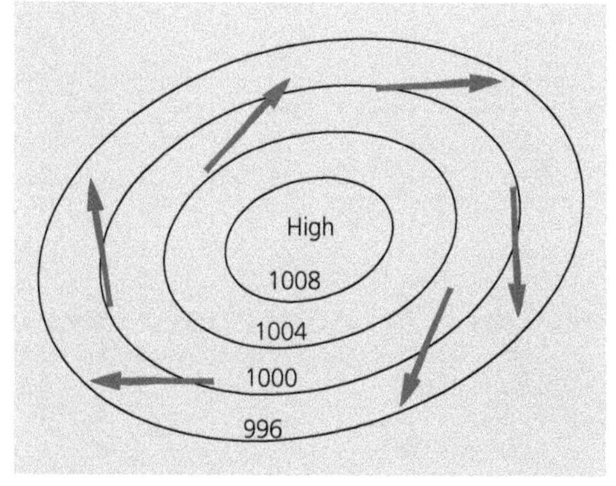

Figure 25 Plan view of an anticyclone.

Figure 26 The weather in a summer anticyclone and a winter anticyclone.

Activities 

1. What is an anticyclone? Use the terms pressure, sinking air, condensation and weather in your detailed definition.
2. Describe the weather conditions shown in the two photographs in Figure 26.
3. Explain why summer anticyclones bring warm weather and winter anticyclones bring cold weather.
4. Suggest three ways a large supermarket might adjust its shop stock during a summer anticyclone. Explain your answers.

By the end of this section you will be able to:

- read synoptic charts
- read satellite images
- understand the limitations of weather forecasting in terms of accuracy and range.

Interpreting synoptic charts and satellite images

What are synoptic charts?

Synoptic charts are maps which summarise the weather conditions at a particular point in time for an area. They record the weather using a set of symbols. They show the fronts of a depression and the variation in the pressure of the air using isobars.

The weather at a weather station is represented by these symbols. Each symbol tells you something important about the weather at that station. The symbols are read as shown in Figure 28.

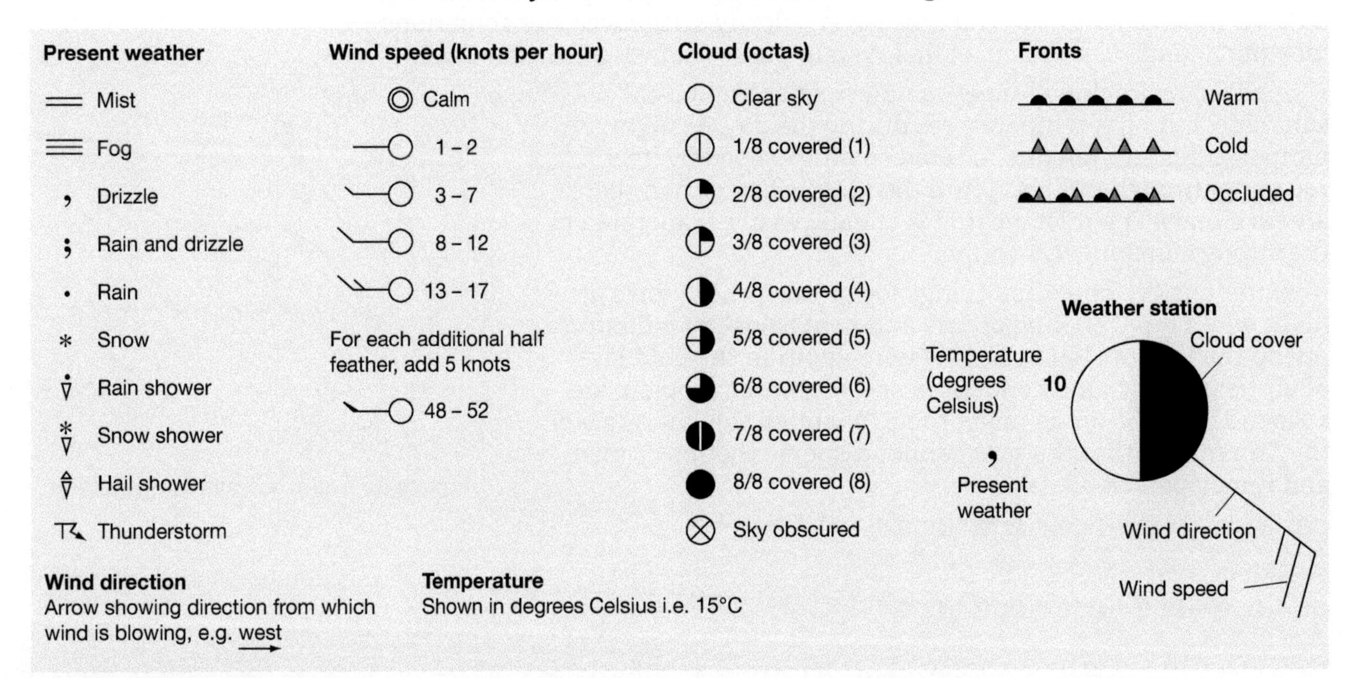

Figure 27 Weather symbols used on weather maps.

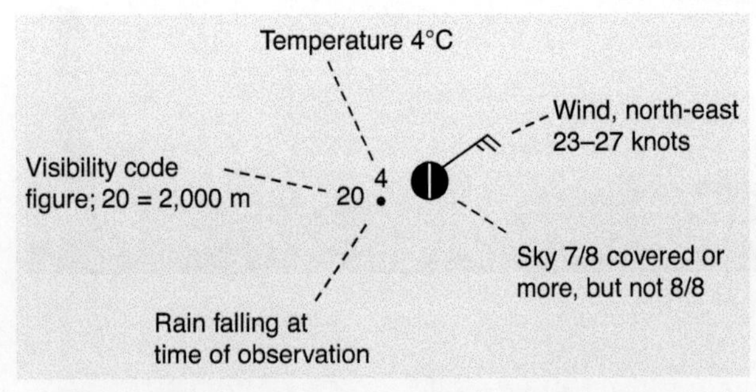

Figure 28 How to read a weather station symbol.

Activities

1. What is the weather like at this weather station? Give precise details of:
   - temperature (in °C)
   - wind speed (in knots)
   - wind direction
   - type of precipitation
   - amount of cloud cover (in oktas).
2. What is wrong with this weather? Why couldn't this happen?

Tip

You only need to be able to read synoptic charts, so don't waste time learning all the symbols: a key of these will be given to you in the examination.

Activities 

The synoptic charts and satellite images below show the development of a depression.

Stage 1: origin and infancy

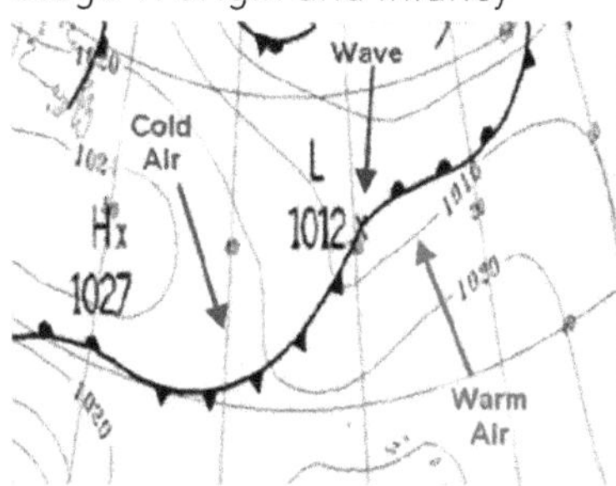

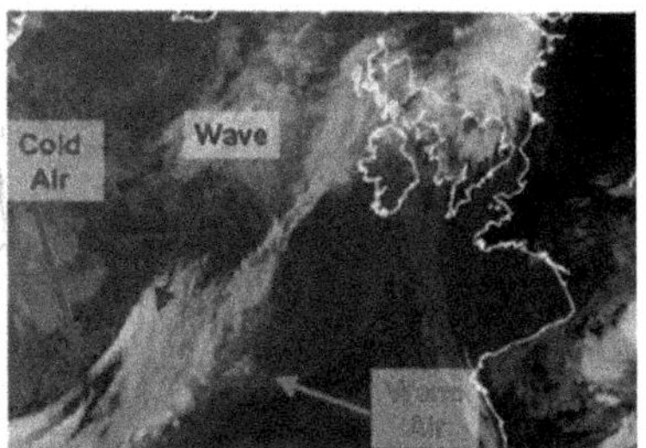

Stage 2: maturity

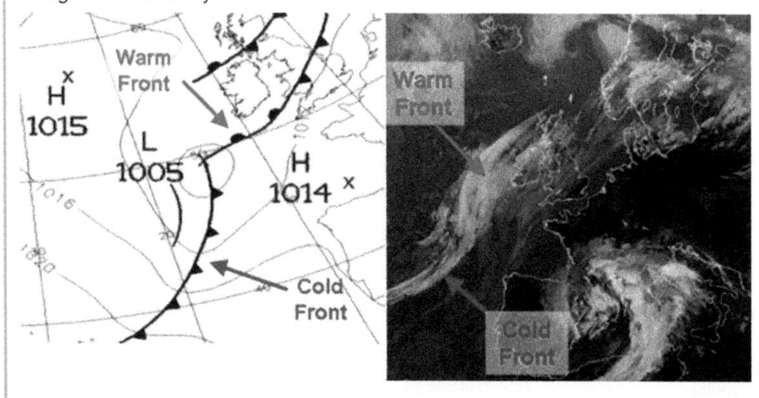

Stage 3: occlusion

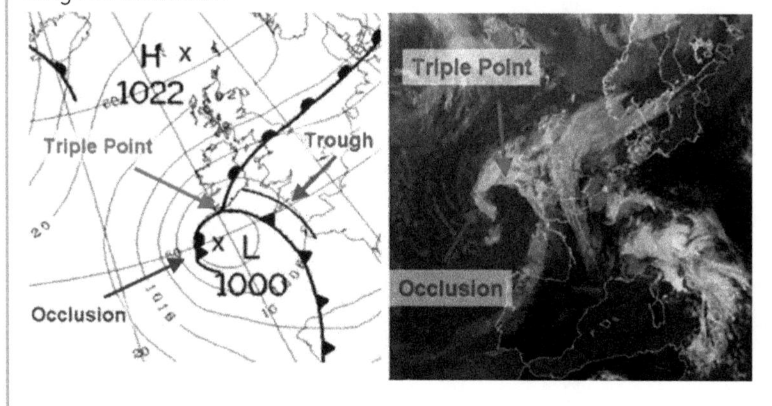

Stage 4: death

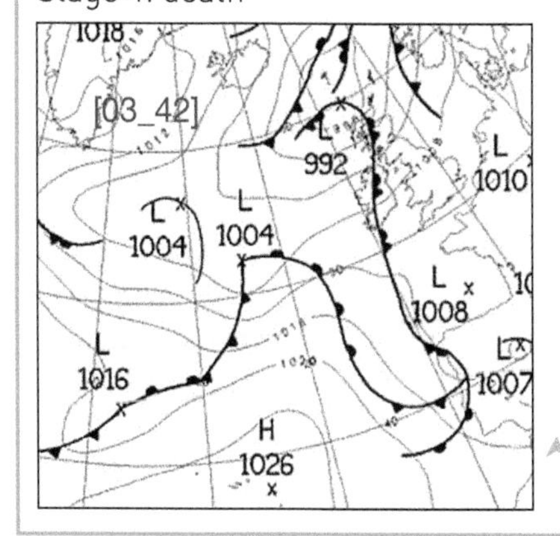

Figure 29 The development of a depression. Source: Met Office.

1 Describe what is happening at each of the four stages in the development of a depression.
2 a Using Figure 22 on page 67, describe how the temperatures change at Belfast through the day.
 b Explain why temperatures change as the depression passes.
 c Describe how the wind direction changes at Derry through the day.
 d Explain why the rainfall amount changes during the day.

Weblinks

www.metoffice.gov.uk/

www.met.ie/

www.accuweather.com/ukie/index.asp?

www.bbc.co.uk/northernireland/schools/

www.weather.gov/

Activities

The synoptic chart in Figure 30 shows an anticyclone affecting the UK.

1. Describe the weather conditions that will be experienced in Belfast if this was from January. Make sure you include temperature, general weather conditions and cloud cover.
2. Describe the weather conditions that will be experienced in Belfast if this was from July. Make sure you include temperature, general weather conditions and cloud cover.
3. Explain why these differences might occur.

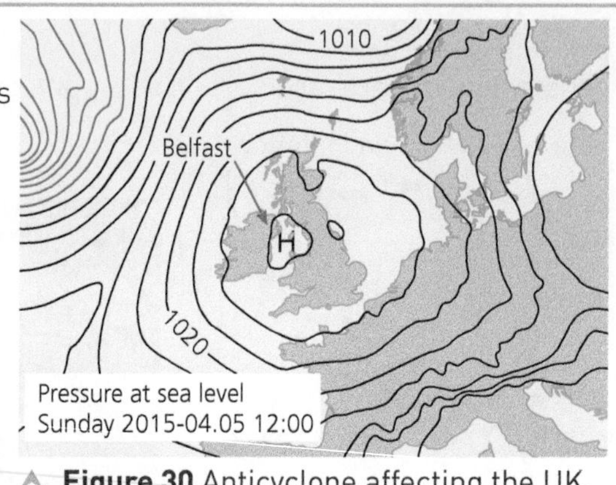

Figure 30 Anticyclone affecting the UK.

What are satellite images?

A satellite image is a photograph taken from space and sent back to Earth. It can show the cloud formations and the pattern of clouds at fronts in depressions, or the clear skies associated with high-pressure areas.

Figure 31 Satellite image of Storm Doris from the Met Office.

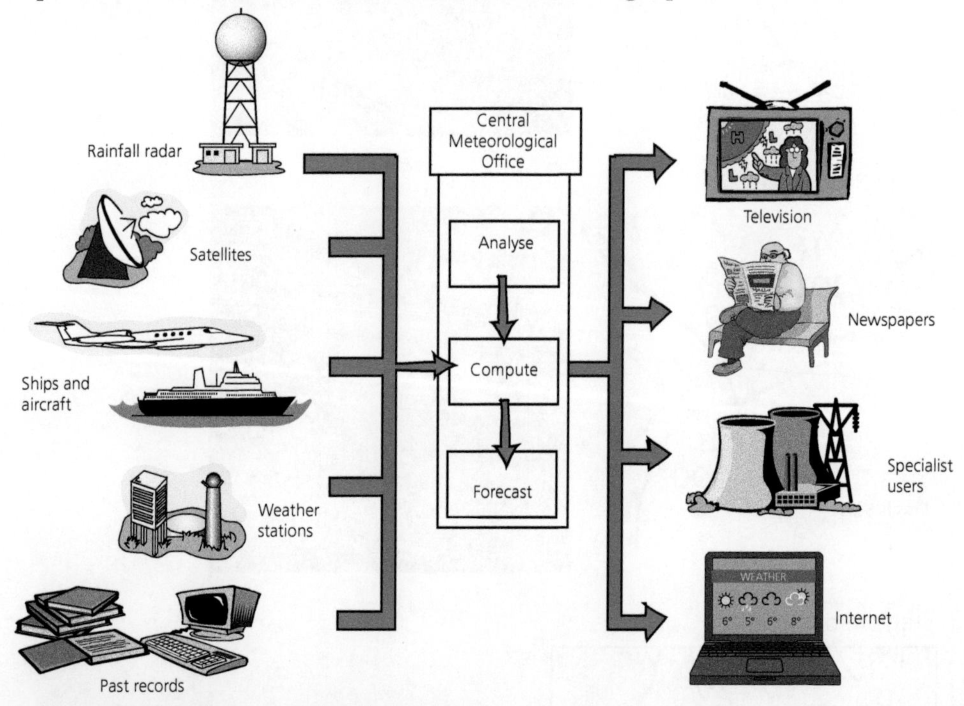

Figure 32 Making a weather forecast.

Activities

1. Study the satellite image in Figure 31. How would you describe the weather being experienced by the UK and Ireland at this time?
2. On a copy of the satellite image, mark the fronts (warm, cold or occluded) and the centres of areas of high or low pressure.
3. Add a suitable key to your satellite image.
4. Using the evidence from the satellite image, explain the spatial variation, or lack of spatial variation, in the weather over the UK on this day.

What are the limitations of weather forecasting?

The range of a forecast refers to how far ahead the weather predictions are in time, whereas accuracy is how successful/reliable the weather predictions prove to be.

Short-term weather forecasting uses weather data collected from all over the world by weather stations, ships, drones and satellites. This information is put into a supercomputer, which tries to work out the most likely changes. As computer systems become more powerful and the weather observations more detailed, the predictions are becoming more accurate, even allowing hour-by-hour predictions for 24–36 hours ahead. However, meteorologists recognise that weather forecasts only have reliable accuracy for a period of up to five days ahead.

Accurate long-term weather forecasting is a future possibility, but is not a current reality. This is because of the chaotic nature of our atmosphere. Chaos theory was introduced to the scientific community in 1972 by Edward Lorenz, when he presented a seminar to academics called 'Does the Flap of a Butterfly's Wings in Brazil set off a Tornado in Texas?' The basis of chaos theory is that the natural system has so many variables that can change the outcome of events that it is unpredictable. Even something as minor as a butterfly flapping its wings in one country could alter conditions enough to cause a tornado somewhere else.

Meteorologists are trying to include chaos theory into computer programs that run analysis of weather data to produce a long-term weather forecast. They alter many common variables to see which outcomes are the most common and release that as a long-term forecast with a reliability percentage attached. For now, that is the best we can expect, but with future advances in computer capacity, even quantum computers, more accurate long-term weather forecasts may be possible.

Activities

1. Using the internet, copy out a weather forecast for your area for the next 24 hours. Comment on whether or not it was accurate – use visual observations of cloud cover, wind speed and rainfall. If you are in a car with a temperature reading, use that too.
2. Choose two of the groups in Figure 33 and suggest why they might need to know the forecast you wrote about.
3. Explain why long-term weather forecasting is less accurate than short-term forecasting.

Weblinks

www.youtube.com/watch?v=fdErsR8_NaU – A video showing how we forecast the weather.

www.youtube.com/watch?v=OoyQVDqbun0& – A video showing the limitations of weather forecasting.

Activities

Look at the following list of terms related to weather:

1. rising air
2. rainfall
3. depression
4. stratus cloud
5. clear skies
6. fronts
7. falling air
8. calm
9. warm sector
10. strong winds
11. frost
12. anticyclone
13. frontal rainfall
14. high temperatures
15. single air mass

1. With a partner write out the sets of terms the following numbers refer to.
   - Set A: 9 14 15 12
   - Set B: 13 9 7 3
   - Set C: 1 8 5 12
   - Set D: 10 13 4 11
2. Work out which one is the odd one out, and explain your reason.
3. Try to add in a new word that connects with the other three terms.

Figure 33 Groups of people who rely heavily on weather forecasts.

4 The impacts of extreme weather on people and property

By the end of this section you will be able to:

- describe the impacts of an extreme weather event that happened outside the British Isles.

Tip

Make sure you learn a selection of impacts, both on people and property, in relation to the case study.

Thankfully, the weather of the British Isles is seldom extreme. However, other parts of the world experience devastating tornadoes, droughts and hurricanes. The bar graph in Figure 34 shows how deadly such events can be even for more economically developed countries like the USA. For poorer countries, the death rate can be much higher. For example, 2,000 people die in India each year from lightning strikes alone.

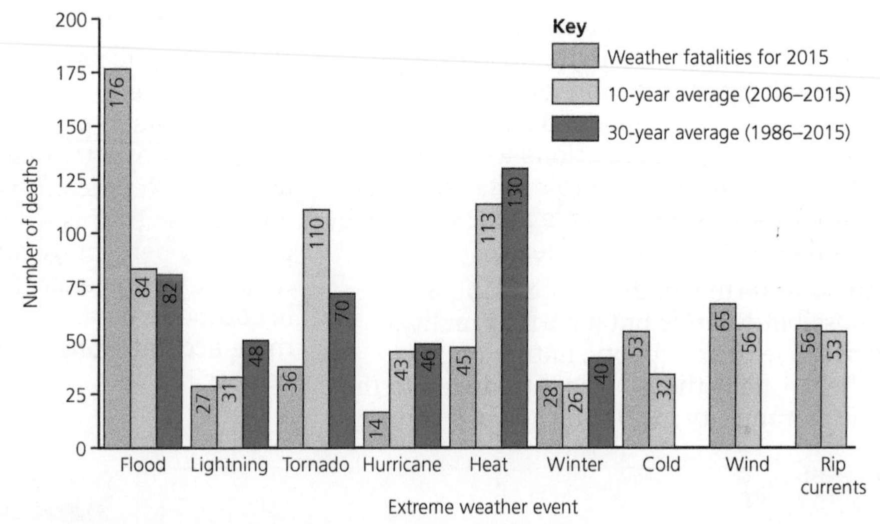

Figure 34 Deaths caused by extreme weather in the USA.

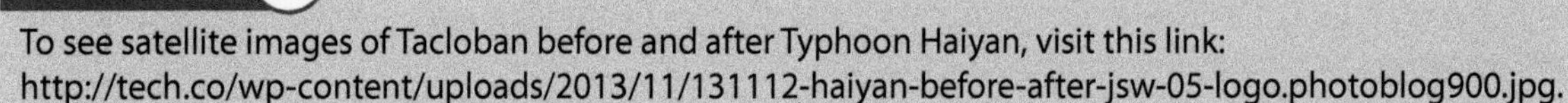

Weblinks

To see satellite images of Tacloban before and after Typhoon Haiyan, visit this link: http://tech.co/wp-content/uploads/2013/11/131112-haiyan-before-after-jsw-05-logo.photoblog900.jpg.

CASE STUDY

Typhoon Haiyan, 2013

- Typhoons are tropical storms created from low pressure systems in tropical areas. They are also known as hurricanes or cyclones. They have a distinct structure with spiralling bands of cloud and a central 'eye' which is free of cloud. They are very large and can easily be seen from space.
- The Philippines is particularly vulnerable to typhoons because of the large amount of warm water that surrounds the 7,000 islands that make up the country. Evaporation of this water followed by condensation releases huge amounts of energy as latent heat that is found within typhoons. There are few pieces of land to slow the typhoons down, so they can become very powerful storms.

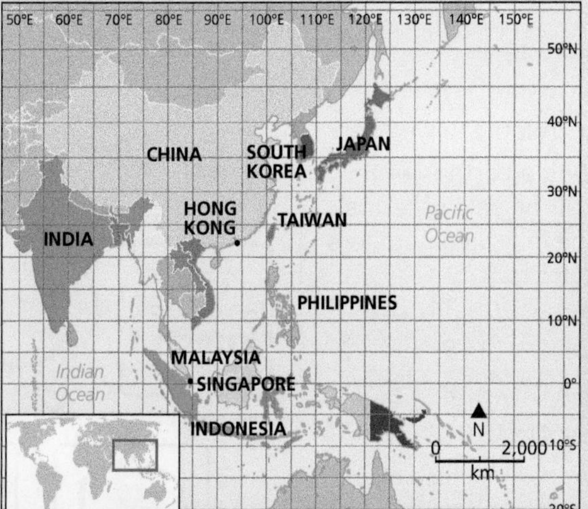

Figure 35 Maps showing the location of the Philippines.

Most typhoons (known as cyclones or hurricanes elsewhere) do not hit land at their peak, but Haiyan did.

Background

- Typhoon Haiyan was the most powerful hurricane ever to hit this area. Locally it was called Typhoon Yolanda and it made landfall in the Philippines on 8 November 2013. Haiyan was a Category 5 hurricane, meaning it was the most severe (see Figure 36) and it brought sustained winds of 230 km/h and gusts of 280 km/h. In the days before the storm reached land, it was watched carefully by the Joint Typhoon Warning Center (JTWC) and the Philippine Atmospheric, Geophysical and Astronomical Services Administration (PAGASA) using satellites and local weather station data. The Philippines is a country which is prone to disasters, not only to typhoons, but also volcanoes, landslides, floods and earthquakes. Finding safe land on the thousands of islands that makes up the country is an enormous challenge. Typhoon Haiyan tracked through the Philippines in only 24 hours, killing an estimated 6,190 people. It caused a storm surge (a wall of water) – that was 8 metres high in some areas, including in the town of Tacloban. It then moved off across the South China Sea towards Vietnam (see Figure 38).

Figure 37 Satellite image of Typhoon Haiyan.

Figure 38 Track of Typhoon Haiyan.

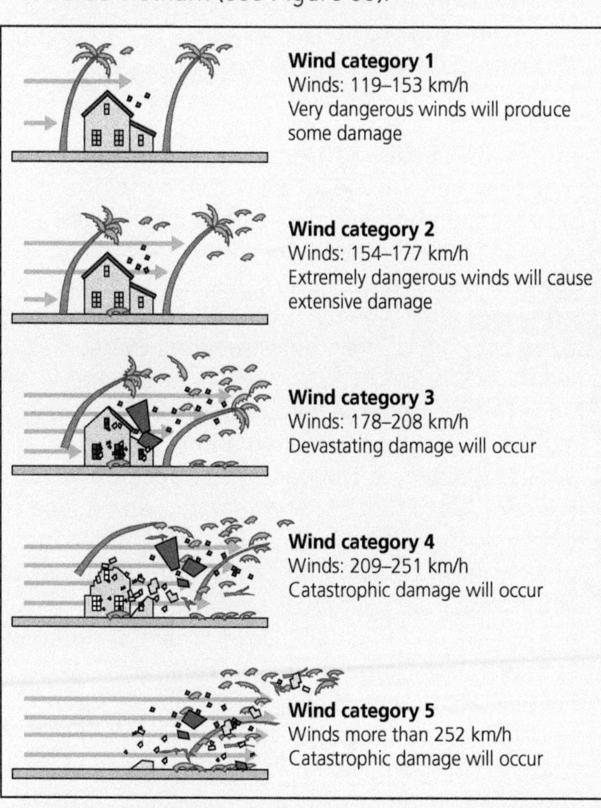

Well-constructed frame homes could have some damage to the roof, shingles, vinyl siding and gutters. Large branches of trees will snap and shallowly rooted trees may be toppled. Extensive damage to power lines and poles will likely result in power outages that could last from a few to several days.

Well-constructed frame homes could sustain major roof and siding damage. Many shallowly rooted trees will be snapped or uprooted and block numerous roads. Near-total power loss is expected with outages that could last from several days to weeks.

Well-built frame homes will incur major damage or removal of roof decking and gable ends. Many trees will be snapped or uprooted, blocking numerous roads. Electricity and water will be unavailable for several days to weeks after the storm passes.

Well-built frame homes can sustain severe damage, with loss of most of the roof structure and/or some exterior walls. Most trees will be snapped or uprooted and power poles downed. Fallen trees and power poles will isolate residential areas. Most of the area will be uninhabitable for weeks or months.

A high percentage of frame homes will be destroyed, with total roof failure and wall collapse. Fallen trees and power poles will isolate residential areas. Power outages will last for weeks to possible months. Most of the area will be uninhabitable for weeks to months.

Figure 36 Saffir–Simpson hurricane wind scale.

What were the impacts of Typhoon Haiyan on people?

- The government of the Philippines said the storm resulted in over 6,190 deaths, and over 1,785 people were reported missing. The city of Tacloban, home to more than 220,000 people, suffered more loss of life than any other area of the Philippines.
- Hundreds of thousands of people were evacuated by authorities before the storm reached land. When Typhoon Haiyan made landfall, 371,000 people were evacuated and many more ended up living in refugee centres which could not withstand the strong winds, and so were themselves destroyed.
- There was also severe damage to crops. Although the harvest season for crops like sugarcane, rice and coconut had just ended, many of the seed stocks were destroyed. This resulted in food shortages for 2.5 million people.
- In the aftermath of the storm in the Philippines there was further loss of life caused by diseases such as cholera and dysentery due to decaying corpses and raw sewage spread by flood waters contaminating drinking water supplies.
- Millions of people experienced disruption to their electricity supply throughout the Philippines. The city of Bogo was blacked out and it took weeks to restore power.

What were the impacts of Typhoon Haiyan on property?

- Being such a strong storm, there was a lot of property damage. Five million people saw their homes severely damaged or destroyed. Ninety per cent of all structures were wiped out across a 500 mile radius from the eye of Typhoon Haiyan. Those made homeless were mainly in the western and eastern Visayas.
- In all 10,390 schools were destroyed, so many children missed out on their education despite attempts to create temporary school structures following the storm.
- Even newly built civic buildings could not withstand the powerful winds and high storm surge. The city of Bogo had just celebrated the opening of its new town hall in April 2013, but Typhoon Haiyan all but destroyed it in November as the roof was ripped off, windows were broken and some walls collapsed. In Tacloban, many people sought refuge in an indoor stadium. It had a reinforced roof to withstand strong winds, but they died when the structure was flooded.

Figure 39 Tacloban after the typhoon.

- The main airport in Tacloban was severely damaged. The terminal building of Tacloban Airport was destroyed by a 5.2 metre storm surge, which reached up to the height of the second storey. The runways were submerged by water and could not be used for landing or take-off.

Conclusion

- Immediately following Typhoon Haiyan, it was difficult to get emergency food, water and medical supplies to people as the airport was underwater and trees had blocked many roads. However, Tacloban airport managed to reopen three days after the storm, so aid began to arrive to the city. The US navy delivered fresh water and the UN released $25 million in emergency funds to provide immediate assistance. This was increased as celebrities like the Beckhams and large multinational companies like Coca Cola and Walmart raised money for emergency assistance to the country. Within two weeks, over 1 million food packs and 250,000 litres of water had been distributed.
- In July 2014, the Philippine government declared its 'Build Back Better' plan for the nation. This included a no-build zone along the coast of the eastern Visayas, a new storm-surge warning system, mangrove-replanting schemes and plans to build an embankment to protect low-lying parts of Tacloban.

- The typhoon's impact on human life was great because of where it hit. Due to widespread poverty and low economic development, many Filipinos live in wooden homes, which are unlikely to survive intense winds. Many storm shelters didn't survive the storm due to poor construction and, being a set of small islands, so many areas were close to the sea and offered few real evacuation possibilities. Such factors contributed to the large scale of this disaster. However, even well-built structures cannot fully withstand a Category 5 storm.
- The Philippines Red Cross delivered basic food aid, which includes rice, canned goods, sugar, salt, cooking oil. The organisation also released a set of guidelines for how to survive in the aftermath of the storm.

The Philippines Red Cross 'Survival Tips'

- If your house was destroyed, make sure that it is safe before you try to retrieve belongings.
- Beware of dangerous animals such as snakes that may have entered your house.
- Watch out for live wires or sockets immersed in water.
- Report damaged electrical cables and fallen electric posts to the authorities.
- Do not let water accumulate in tyres, cans or pots to avoid creating a favourable condition for mosquito breeding.

Activities

1. Describe the graph on page 74 showing weather fatalities in the USA.
2. Suggest why the Philippines is described as a disaster-prone country.
3. List the three main dangers of typhoons.
4. Why did the Red Cross advise against allowing standing water to accumulate around people's homes?
5. Summarise the information given in this case study. Use this layout:
   - title, name, location and date of extreme weather event
   - background information on the weather event
   - impacts of the event on people
   - impacts of the event on property
   - conclusion on overall deadliness of this weather event.

Figure 40 One family's story.

They spent the first few weeks after the typhoon hunting for food and their two small children. They walked through the rubble of their city, among the dead and the living, like zombies, asking neighbours and scanning ad hoc burial sites. They even waded into the oily waters where their house once stood – before the ships pummelled their waterfront neighbourhood, leaving a mess of bloated bodies, twisted metal and broken concrete.

It was one of those ships that saved Urwin Coquilla, 37, and his wife, Ethel, that early grey morning of 8 November. Typhoon Haiyan, the strongest storm to make landfall in recorded history, tore through the Philippines at speeds of 195mph, ripping up farms, levelling villages and leaving more than 6,300 dead. The family hadn't evacuated their home in Tacloban – Ethel believed they would weather the storm, as they had every other – so when the tidal wave shattered the floor from underneath their feet, and the ships came rolling across the land, they were swept out into the swirling black water. The children clung on desperately to their parents, but they were torn from Urwin's arms and disappeared into the storm. There was so much in the water – debris, people, animals, filth – that when a massive ship hurtled towards him, Urwin could only just grab the rope hanging from its deck and fling his wife aboard. Then, shakily and clumsily, he climbed on as well, only to fight off the ship's enraged captain, who was screaming, through the howling wind, that the couple weren't authorised to use the life jackets hanging in the storeroom.

When the waves receded, Urwin looted destroyed shops for supplies and searched under piles of rubble for his children, but never finding them. Even after they began rebuilding their home and spent their days queueing for relief goods, still the couple searched.

To give up felt impossible, Urwin says, now a bald man in ill health, his wife pale and thin. The toll of the aftermath of Haiyan was clear to be seen. However, after weeks for searching 'we decided to accept the fact that our children were gone, we had to move on and continue living.' Like so many others who lost everything to the storm – their home, their family, their livelihoods – he shrugs at the question. How does a person, a city, a whole country even, recover from such a horrendous event?

Sample examination questions

1 State the difference between 'weather' and 'climate'. [3]

2 Study Table 1 and complete it by writing the correct answers in the blank boxes. [3]

Name of weather Instrument	Weather element recorded	Unit of measurement
Barometer	Air pressure	
	Rainfall	mm
Anemometer		Knots

Table 1 Weather instruments.

3 Describe how to use a digital thermometer. [3]

4 In the box below, underline the cloud type which is found in the highest sections of the atmosphere. [1]

Cirrus Cumulus Stratus

Tip

When explaining something, always elaborate on a point you make rather than creating a list of unrelated facts.

5 Explain how satellites can help to create a weather forecast. [3]

Tip

When describing something from a map, ensure to include distance, direction and place names if possible.

6 Describe the distribution of rainfall shown on the rainfall radar image in Figure 1 on page 79. [3]

7 Study Figure 2 on page 79, which shows the main factors affecting climate. Choose two of the factors and explain how each of them affects climate. [6]

Figure 1 Distribution of rainfall in the 72-hour period from 9 a.m. on 17 November to 9 a.m. on 20 November 2009.

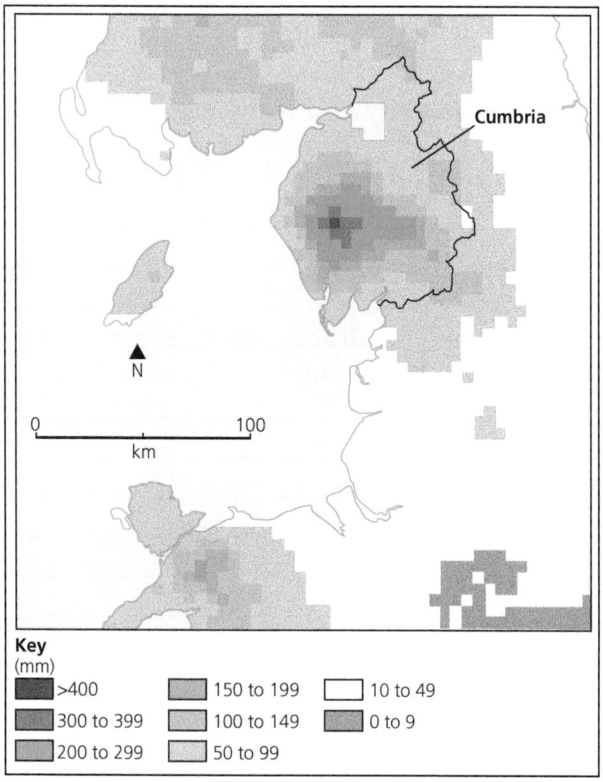

Figure 2 The main factors affecting climate.

Figure 3 The weather chart for 6 a.m. Thursday 19 November 2009, with a deep Atlantic depression.

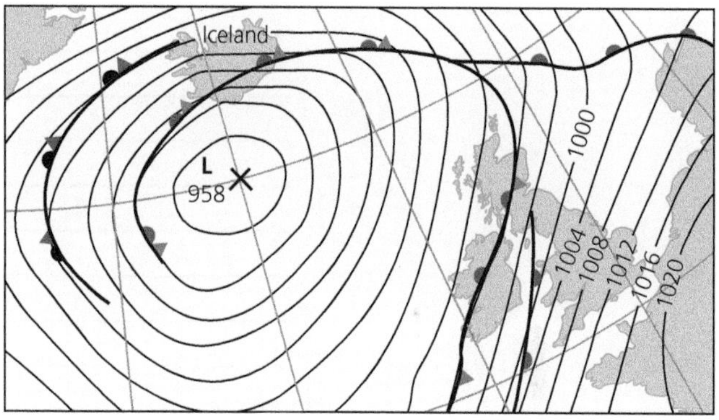

8 State the meaning of the term **air mass**. [2]

9 Compare the characteristics of a Tropical Maritime air mass and a Polar Continental air mass. [3]

Tip

When comparing, ensure you use comparative adjectives such as 'colder' or 'warmer'.

10 Study Figure 3 above, which shows a synoptic chart for Thursday 19 November 2009. Answer the following questions.

(i) State the pressure at the centre of this depression. [2]

(ii) Name the type of weather front over Iceland. [1]

(iii) Explain why depressions such as this one bring wet and windy weather. [5]

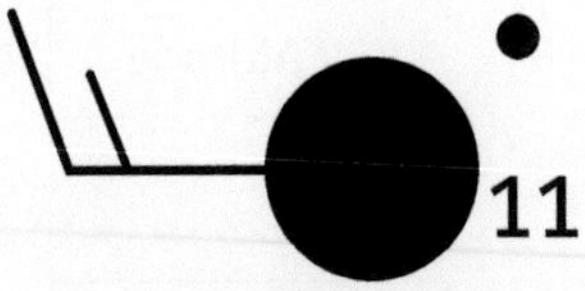

____

____

____

____

11

Figure 4 Weather station data, Belfast, 21 February 2017.

11 Study Figure 4, which shows the weather station information for Belfast on 21 February 2017. Use it and the key on page 70 to help you complete the following weather report for that day:

Today there will be ***full / no*** *cloud cover over Belfast.*

There will be ***rain / sunshine*** *and a* **Westerly / Northerly** *wind blowing at* *10 / 15 knots.* [4]

12 State the meaning of the term 'anticyclone'. [2]

____

____

____

13 Compare and contrast the weather typically experienced under a winter anticyclone and a summer anticyclone. [6]

____

____

____

____

____

____

____

____

____

____

____

____

14 Describe the impact on people and the environment of your chosen example of an extreme weather event from outside the British Isles. [6]

____

____

____

____

____

____

____

____

____

____

____

____

UNIT 1

UNDERSTANDING OUR NATURAL WORLD

THEME D: The Restless Earth

▲ Damage caused by a tsunami in Japan.

What other kind of damage can be caused by earthquakes?

1 Plate tectonics theory

By the end of this section you will be able to:

- describe the structure of the Earth
- explain how the Earth's crust is made up of a number of plates
- explain how convection currents in the mantle cause plate movement.

Tip 

Think carefully about the two types of crust. Having a clear understanding of how oceanic and continental crusts behave will make it easier to explain the landforms and hazards seen where these two types of crust meet.

The structure of the Earth and tectonic plate movement

What is the structure of the Earth?

The Earth is made up of a series of layers.

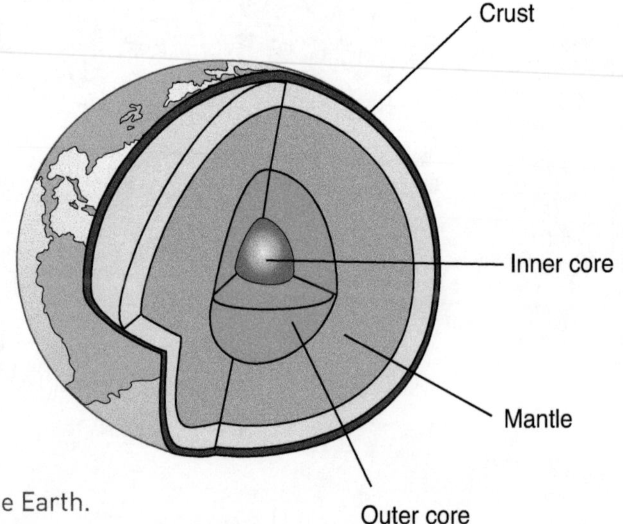

Figure 1 Layers of the Earth.

The bit of the Earth you are on is moving at 70 mm per year. The Earth is revolving round the Sun at 30 km per second. Our solar system is revolving around the centre of the galaxy at 300 km per second and even our galaxy is moving through the universe at 600 km per second! Astonishing speeds!

The centre of our planet is called the core. There are two distinct layers within the core. The outer core is fluid and mostly made of iron and nickle. The inner core is solid. Beyond the outer core is the mantle. This layer is almost 3,000 km thick and accounts for 84% of the earth's mass. It is composed of very hot, dense rock.

There are two main types of crust: oceanic and continental.

The table below summarises the characteristics of the two different types of crust.

Type of crust	Density	Thickness	Behaviour	Strength	Age of crustal material
Oceanic crust	Very dense (heavy) Mean density is 3,000 kg/m^3	Thin: 5–10 km	Can sink into the mantle	Easily destroyed	Young crustal material
Continental crust	Less dense (light) Mean density is 2,700 kg/m^3	Thick: 30–70 km	Does not sink easily into the mantle	Hard to destroy	Old crustal material

Is the Earth's crust solid?

Plate tectonics theory proposes that the Earth's crust is split into sections called tectonic plates. These plates are constantly moving around on top of the mantle, at an average speed of 70 mm per year. The places where plates meet, called plate margins, are related to seismic (earthquake) and volcanic activity.

The consequences of such movements of landmasses and the opening up of new oceans and seas are far reaching. The process has influenced climate change and even the spread of plants and animals. The opening of the Atlantic Ocean (between 80 and 65 million years ago) created a new ocean gateway for heat transfer between the tropics and the poles, changing global temperatures. In addition, the breaking away of Australia from a central landmass allowed the evolution of marsupial animals like the kangaroo, which are not found on other continents.

Tip 

You do not need to learn the names of all the plates, but it can help if you are familiar with them as this can save you time when answering examination questions.

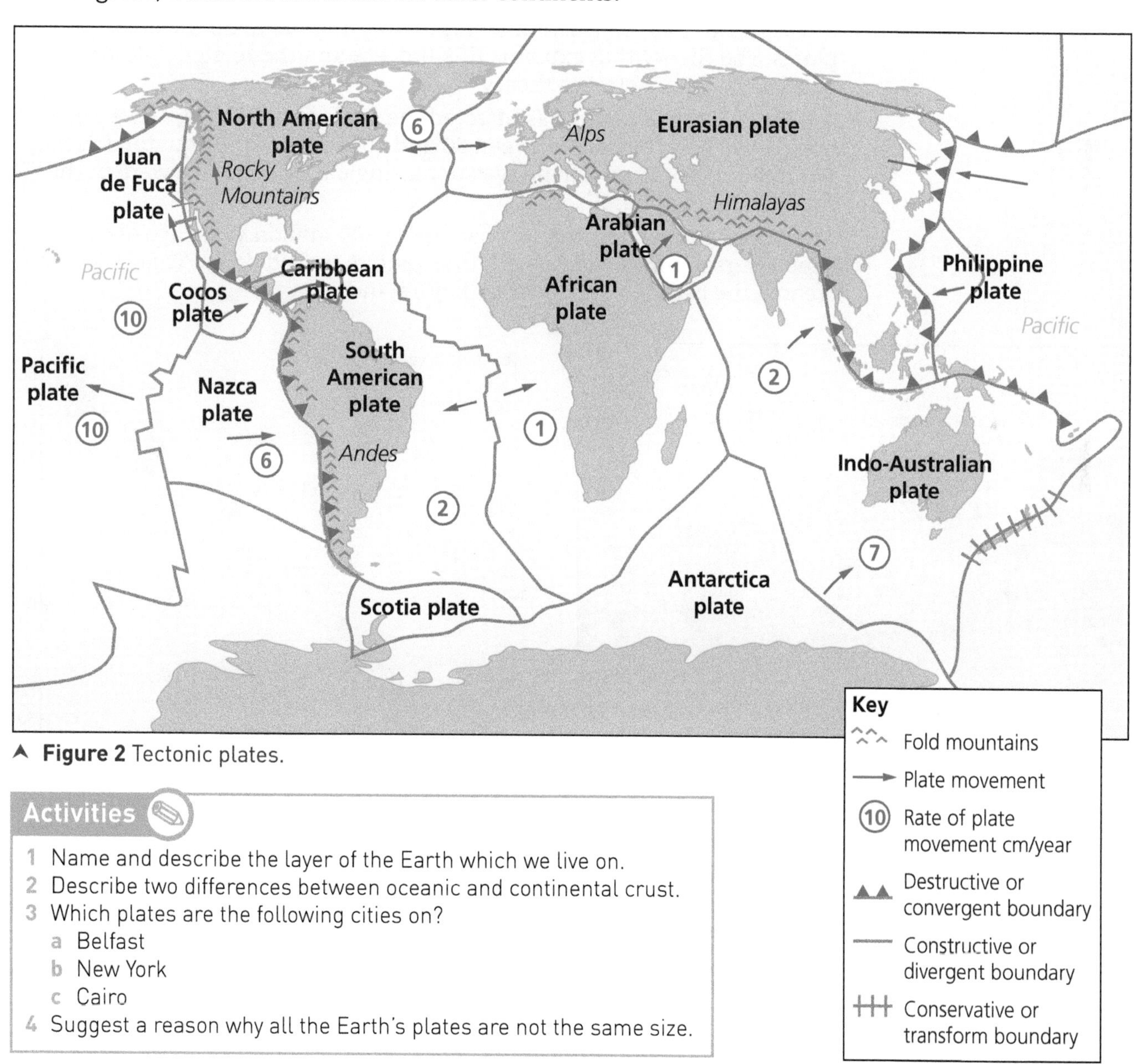

Figure 2 Tectonic plates.

Activities

1. Name and describe the layer of the Earth which we live on.
2. Describe two differences between oceanic and continental crust.
3. Which plates are the following cities on?
   a. Belfast
   b. New York
   c. Cairo
4. Suggest a reason why all the Earth's plates are not the same size.

Tip

The impact of convection currents and the plate movements they cause can be seen clearly if you go to www.youtube.com/watch?v=uLahVJNnoZ4. Other video clips show how the positions of the plates have changed over time; go to www.youtube.com/watch?v=1-HwPR_4mP4.

What makes plates move?

In a more detailed cross-section of the Earth (Figure 3) it can be seen that the crust on which we live acts as though it is floating on a layer of molten material; this is called the mantle. Inside the Earth there are convection currents within the mantle that move heated molten material upwards from the core, up towards the crust. Here it cools and sinks back down to the core, so that the cycle can start again. It is these convection currents that cause the crust above them to move.

Where currents descend (go downwards) they drag crust into the mantle, creating a destructive plate margin. Here crust is destroyed as it descends down into the hot mantle, where it melts. Where currents ascend (rise up) they pull the crust apart, creating a constructive plate margin. Molten material from the mantle rises to plug the gap in the crust, creating new crust.

It was these same convection currents that first broke up the crust, creating plates. The theory that explains this process and the related landforms and hazards is plate tectonics theory.

All this moving of plates has affected the geology of Ireland, as we have not always been located between 50 and 60° north of the equator. Indeed, 500 million years ago Ireland was not a single island and we were in the southern hemisphere!

The tropical ocean location we had about 300 million years ago allowed deep layers of marine clays and limestone to be laid down. This limestone created the bedrock of almost half of Ireland.

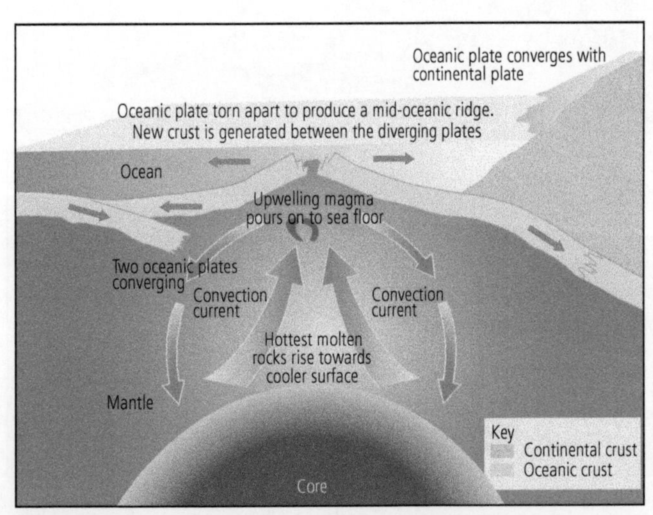

▲ **Figure 3** Convection currents in the mantle.

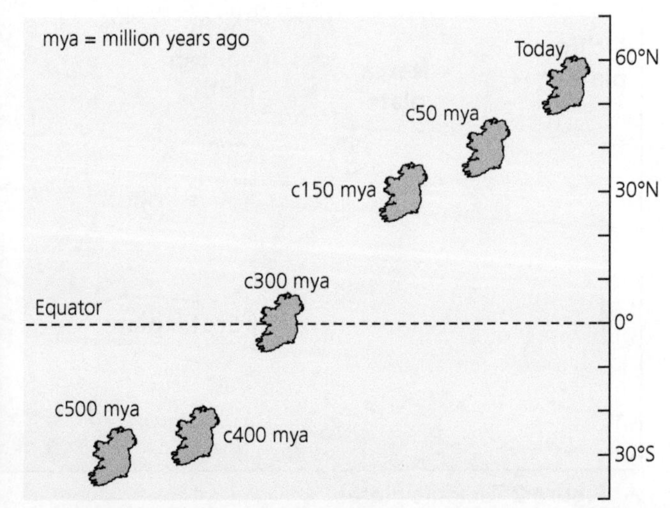

▲ **Figure 4** The changing position of Ireland over the past 500 million years caused by plate movement.

Activities

1. Explain why plates move – in your answer make sure you use the following terms: crust, mantle, core, convection current, heat, movement.
2. Using Figure 4, describe the changing latitude of Ireland over the last 500 million years.
3. Suggest why fossils of tropical plants can be seen in Irish sedimentary rocks.

Plate margins and their processes and landforms

By the end of this section you will:

- know and understand the processes and landforms associated with constructive, destructive, collision and conservative plate margins.

What are plate margins?

There are four main types of plate margin:

- Constructive: where plates are pushed apart, so they move away from one another and new crust is created.
- Destructive: where an oceanic plate crashes into another plate and oceanic crust is pushed into the mantle and destroyed.
- Collision: where two sections of continental crust crash into each other and both are pushed upwards to form a vast mountain range.
- Conservative: where plates slide past each other. Crust is neither created nor destroyed.

Earthquakes and volcanoes are the hazards most often found at plate margins (see sections 3 and 4 for more information). The map below shows the major active zones around the world.

Tip

There is a lot of information about each plate margin, so try to break it down into a mind map for each type.

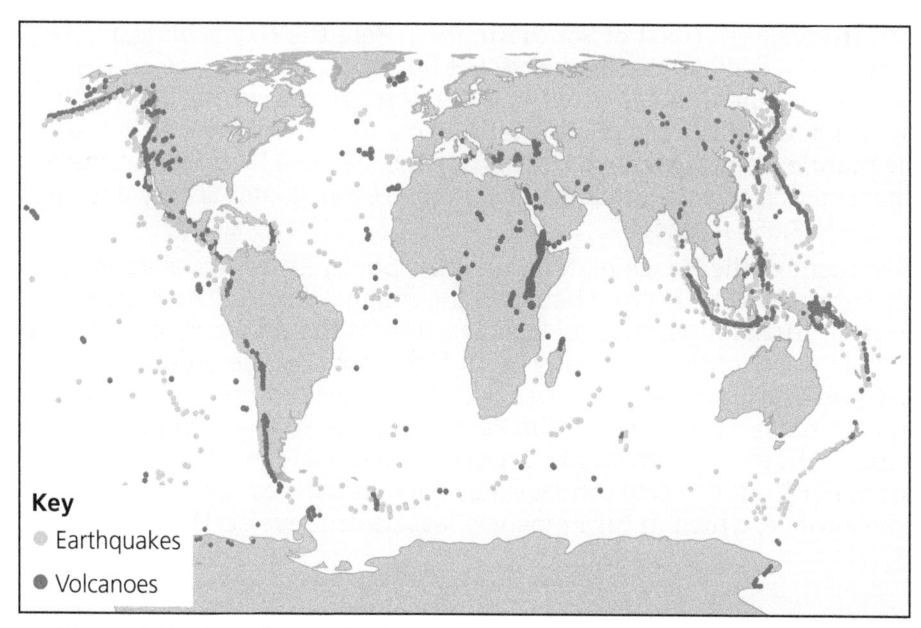

▲ **Figure 5** Earthquakes and volcanoes along tectonic plate margins.

Activities

1. What is a plate margin?
2. Name and describe the four main types of plate margin. Draw simple diagrams to aid your descriptions.

▲ **Figure 6** The San Andreas Fault line.

What are the processes and landforms at constructive margins?

One constructive margin is found in the middle of the Atlantic Ocean. Here the Eurasian and North American plates are being pulled apart, moving away from one another. This means the Atlantic Ocean is getting wider by about 3 cm a year. This movement causes regular, but weak, earthquake activity. Magma wells up from the mantle to plug the gap, so there is often frequent gentle volcanic activity even here under the ocean. This rising of material pushes up the crust slightly at either side of the plate margin, thus creating a mid-oceanic ridge.

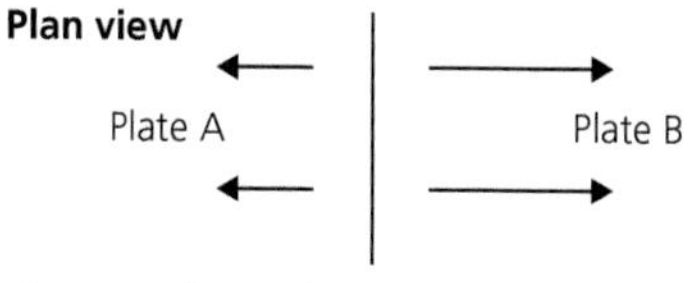

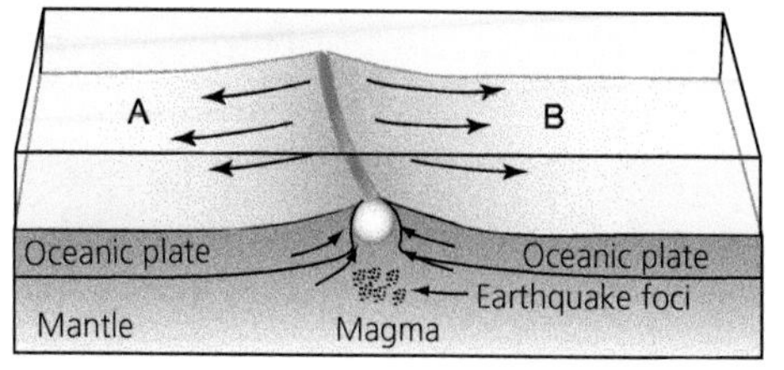

▲ **Figure 7** Constructive (spreading) margin.

Activities

1. Name the main landform feature found at constructive margins that are under the sea.
2. Explain what hydrothermal vents are and why they are found at constructive boundaries.
3. Research either the Mid-Atlantic Ridge or the Great Rift Valley. Use it to draw an annotated diagram of features found there and a fact file about the area.

The volcanoes found along the ridge are called smokers. This chain of volcanic mountains is the longest in the world and means that the middle of the Atlantic is relatively shallow. The hardened lava erupted from volcanoes forms new crust.

Constructive margins are also found where continental crust is splitting apart. Rift valleys are seen at constructive margins on continental crust. One example is the Great Rift Valley in eastern Africa.

What are the processes and landforms at destructive margins?

Destructive margins have a subduction zone. Here oceanic crustal material is being pulled into the mantle, where it melts and is destroyed.

This type of margin falls into two types:

1 Oceanic crust crashing into continental crust.

2 Oceanic crust crashing into oceanic crust.

Oceanic crust–continental crust margin

An example of where oceanic crust is crashing into continental crust can be found on the western coast of South America. Here the Nazca plate, made of oceanic crust, is disappearing below the South American plate. At the plate margin, the heavy oceanic crust is being pushed downwards into the mantle. It is dense (heavy), and so it falls below its normal level as it sinks into the mantle, creating a deep ocean trench called the Peru–Chile trench. This linear trench follows the line of the western coast, and is very deep in places – up to 8,050 m deep!

The movement of the Nazca plate against the South American plate is not smooth, because of the friction between the rough surfaces. The plates may become stuck for years, until the pressure for the plates to move is greater than the friction preventing them from moving. The pressure is released suddenly and the two plates will jolt many centimetres at once. This sudden movement is felt on the Earth's surface as an earthquake. The point at which the earthquake occurs is known as the focus. The deeper the earthquake occurs the weaker the shockwaves are when they reach the surface, which in turn means a less destructive earthquake.

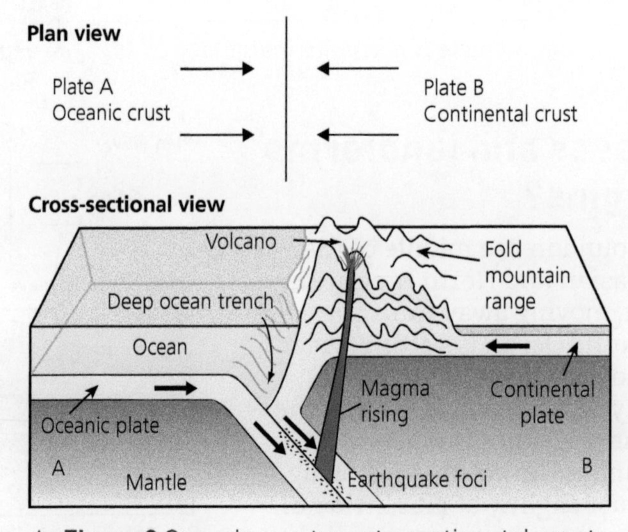

▲ **Figure 8** Oceanic crust meets continental crust.

As the Nazca plate disappears down into the mantle (this is called subduction), it begins to melt due to intense heat. Because it was once oceanic crust, it is saturated with water. The magma created is therefore chemically different from any found naturally in the mantle – it is full of gas bubbles, created by the evaporating water. This means that this melted oceanic crust is less dense (lighter) than the surrounding mantle, so it rises upwards as an explosive type of magma. If it breaks through the surface, it creates a volcano (e.g. Cotopaxi and Chimborazo in Ecuador). The continental crust that makes up the South American plate is not dense like oceanic crust, so it is not subducted easily. Instead it folds and buckles upwards to create a linear fold mountain range, such as the Andes.

Activities 

1. Name one destructive margin where oceanic crust is subducted below continental crust.
2. Explain the connection between subduction of oceanic crust and volcanic activity.
3. Using a diagram to help you, explain why earthquakes with a deep focus are less destructive than those with a shallow focus.

Oceanic crust–oceanic crust margin

An oceanic crust margin, where oceanic crust and oceanic crust meet, has many similar features to the first type of destructive margin. As the oceanic crust sinks into the mantle, it melts and creates a less dense material than the surrounding rock. A deep ocean trench forms where the more dense material is pushed down into the mantle. This oceanic trench can be very deep. For example, the deepest part of the Mariana Ocean trench, in the Pacific, is 11,022 m. This could easily swallow Mount Everest, the tallest mountain on the surface, which is 8,848 m in height. There are no fold mountains as there is no continental crust to buckle upwards.

This magma then rises upwards and may erupt through the crust to create a volcanic island. In this way a chain or arc of such volcanic islands is made. The island arc is aligned to the boundary of where the two plates were moving towards each other.

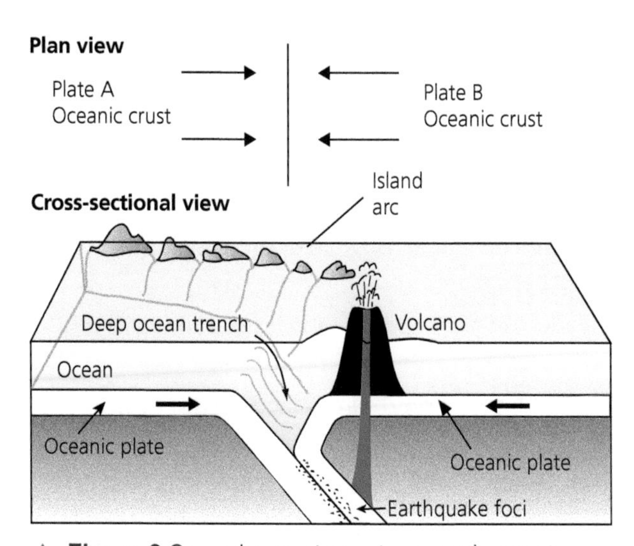

▲ **Figure 9** Oceanic crust meets oceanic crust.

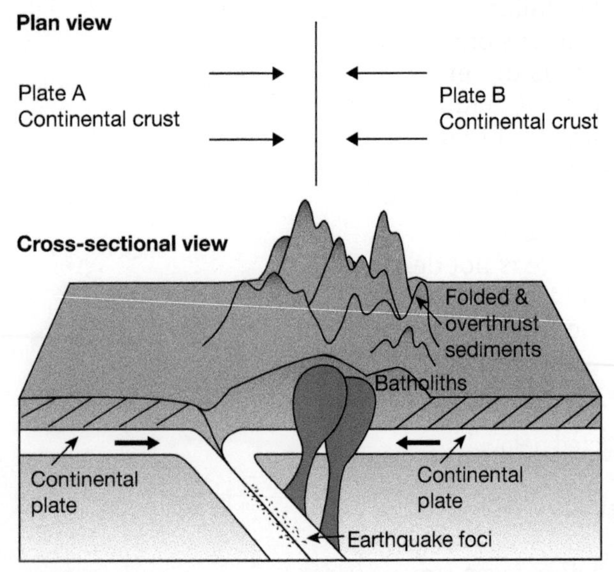

Figure 10 Continental crust meets continental crust.

What are the processes and landforms in collision zones?

Continental crust–continental crust margin

Where two continental plates meet is a collision zone. The crusts of both plates buckle and fold upwards. The two sets of fold mountains overthrust one another, creating a large range of high mountains. There is little material melting, and that which does melt cannot make it through the high mountains to create a volcano. Instead the magma forms large intrusions into the mountain range, called batholiths. The magma cools slowly to form granite cores to the mountains. Examples of mountains formed at this type of margin include the Himalayas and the Andes ranges.

What are the processes and landforms at conservative margins?

At conservative margins, such as the San Andreas fault line in California, two plates try to slide past one another. When friction causes the two plates to stick, pressure to move builds up. This pressure is eventually released as an earthquake when the plates move suddenly. As crust is neither created nor destroyed at conservative margins, there are no volcanic eruptions. Over 20 million people live along the San Andreas fault. Offset streams and irregular lines of trees in orchards are evidence of movement along this conservative margin.

Activities

1. Name the location of one destructive margin where two types of similar crust are moving towards each other.
2. Explain how a mountain range like the Himalayas formed. In your answer make sure you use the following terms: fold mountain, continental plate, buckle, overthrust.
3. Carry out research on the San Andreas Fault. Find an image of the area and write a paragraph to explain why there are so many earthquakes in this area of California.

Weblinks

http://geology.com/plate-tectonics.shtml – an interactive map of some tectonic landforms.

www.geography-site.co.uk/pages/physical/earth/tect.html – the history and rules of plate tectonics.

http://www.weatherwizkids.com/?page_id=98 – a simple introduction with good images.

www.enchantedlearning.com/subjects/astronomy/planets/earth/Continents.shtml – lots of facts and interactive quizzes.

2 Basic rock types

The formation and characteristics of types of rock

By the end of this section you will:

- understand the formation of the main rock types
- be able to recognise the characteristics of the six main rock types.

How are rocks formed?

Rocks are classified by the manner in which they were made, how they have been changed and what they are made from.

Igneous rocks

These are formed when molten lava or magma cools and hardens. If the lava has been exposed on the surface, it may cool quite quickly, producing few if any crystals within its structure; these types of rock are extrusive igneous rocks. The Giant's Causeway in County Antrim is made of one such stone – basalt. Igneous rock can also be made from magma that cooled slowly underground, forming large mineral crystals and speckled igneous rocks; these are intrusive igneous rocks. The granite that is found in the Mourne Mountains in County Down was created this way.

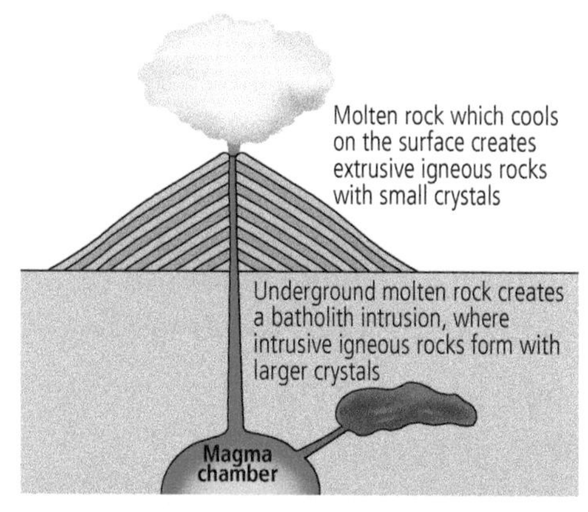

Figure 11 The formation of extrusive and intrusive igneous rocks.

Tip 

To help you remember the different types of igneous rock, use the first letters to help jog your memory: **i**ntrusive form **i**nside the crust, while **ex**trusive **ex**it the crust and are formed on the Earth's surface.

Sedimentary rocks

Weathering and erosion of rocks produces sediments, small fragments or particles, which accumulate on land, coasts and marine environments. Over time, layers of these fragments form on top of one another, causing the air and moisture to be squeezed out, and a solid rock to be created. These are sedimentary rocks. The line between layers of sediment is a line of weakness called a bedding plane.

As well as being made from fragments of rocks, **sedimentary** rocks can be made from the chemicals left after the evaporation of water, or from layers of plant and animal remains. Occasionally, plant and animal remains do not get crushed by the process and remain trapped in the rock as a fossils.

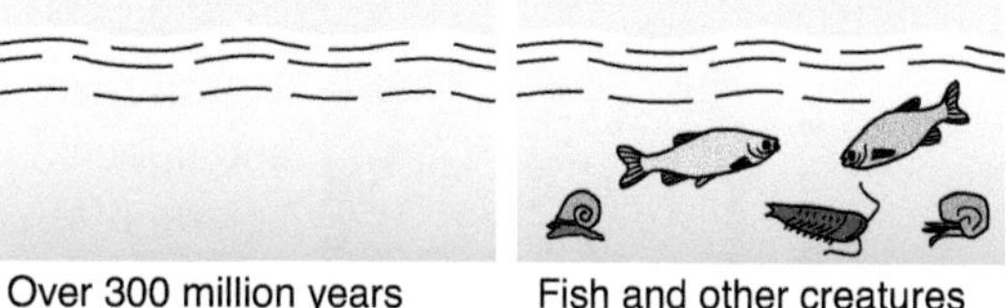

Over 300 million years ago the land that became the UK was under the sea

Fish and other creatures were the only inhabitants

When these creatures died, the skeletons and shells became sediment

Over the millennia, the sediment grew thicker and heavier

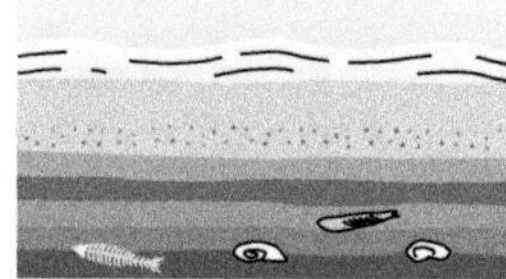

The sediment was compressed and it became limestone

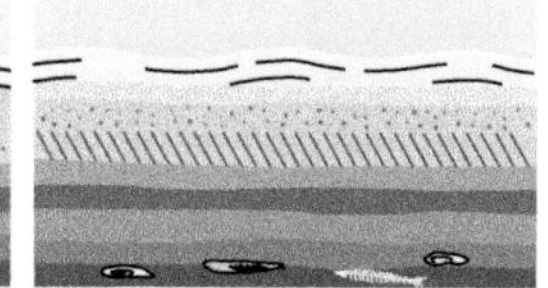

Rivers would carry sand and grit to the sea. When these fell as sediment they eventually became sandstone and gritstone

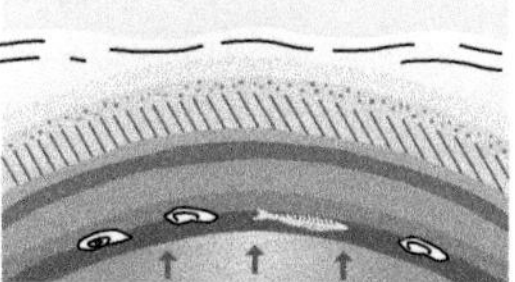

Strong forces in the Earth's crust began to push the seabed upwards

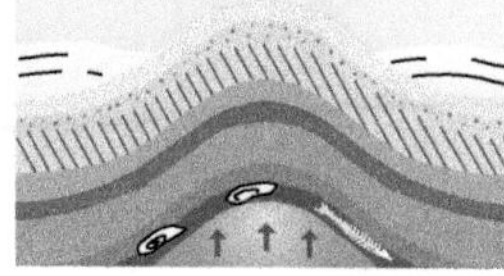

As the sea drained away, the rock that was on the seabed became dry land

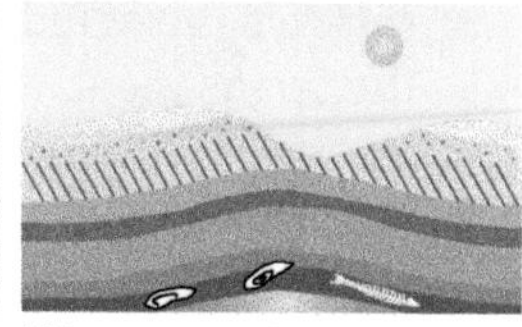

We can see the sedimentary rocks, e.g. limestone and sandstone, that are formed in this way.

Figure 12 How sedimentary rock is formed.

Metamorphic rocks

The final main group of rocks are the metamorphic rocks. These are rocks that have been altered or changed by extreme heat or pressure. They were once either igneous or sedimentary rocks. Sometimes the pressure and heat have changed the rock so much that it can be very hard to tell what it originally was.

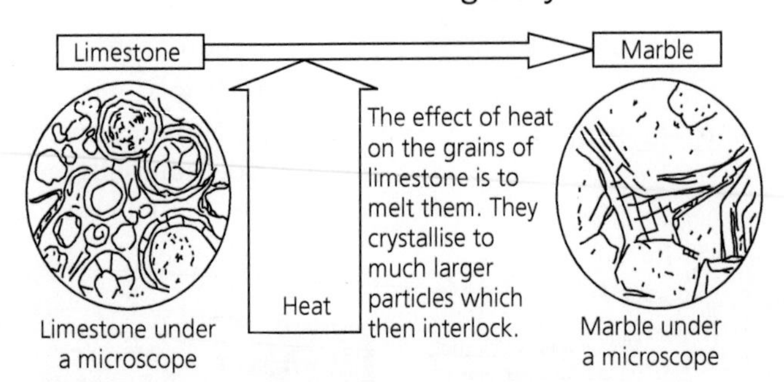

Figure 13 The metamorphosis of limestone into marble.

We can do simple experiments to investigate the origin of metamorphic rocks. Marble is a metamorphic rock that fizzes when acid is poured onto it, just like limestone fizzes under the same experimental conditions. Further chemical analysis of the two types of rock show that they are both mostly made from calcium carbonate. The fragments of shell that created the limestone have been crystallised by heat to make marble.

There are hundreds of different kinds of rocks, but every rock is either igneous, sedimentary or metamorphic. All these rocks are made from pure minerals, but their method of creation is very different.

What are the characteristics of these basic rock types?

This table shows some of the most common rocks for the three rock types. You will need to know their characteristics as you may be asked about them in the examination.

	Rock	Description	Rock	Description
Igneous	**Granite**	Rough texture and speckled colour, often pink or grey.	**Basalt**	Very hard, dark grey rock. Often feels rough and heavy. Small glittery speckles may be visible.
Sedimentary	**Limestone**	Grey, white or yellow rock. May be hard and contain fossils and layers. Porous.	**Sandstone**	Formed from grains of sand. No crystals. Feels rough to touch. Quite hard.
Metamorphic	**Slate**	A dark grey rock with layers which are easily split apart. Smooth, flat surface. It is impermeable because of this.	**Marble**	May be pure white or have swirls or bands of colour running through it. Unpolished, it feels rough and grainy.

Activities

1. Describe what a bedding plane is.
2. Explain the link between metamorphic rocks and the two other rock types; include the terms 'heat' and 'pressure' in your explanation.
3. Create a learning map to help you revise the three types of rock.

The global distribution and features of earthquakes

By the end of this section you will be able to:

- describe the global distribution of earthquakes
- distinguish between the focus and epicentre of an earthquake.

How are earthquakes distributed in relation to plate boundaries?

Earthquakes are the shaking of the ground surface caused by a sudden movement of the Earth's crust. Earthquakes can happen anywhere but, as Figure 14 shows, they are mostly found along three main belts, which consist of the most volatile plate boundaries. These belts:

- encircle the Pacific Ocean
- pass along the coast of the Mediterranean Sea and through southern Asia, towards the Pacific Ocean
- pass through the middle of the Atlantic, Southern and Indian oceans.

Figure 14 also shows the location of some of the most recent earthquakes. Remember, earthquakes are very frequent events. There are millions of earthquakes recorded globally each year. The world's main fold mountains have been formed by the buckling of the Earth's surface during crustal movements felt as earthquakes. There is a clear link between earthquake activity and proximity to plate margins/boundaries – the closer a place is to the edge of a plate, the more likely it is to experience earthquake activity.

Tip

Make sure you can identify the places where earthquakes are common as you might need to name them from an unmarked world map. Look at Figure 14 to help you with this.

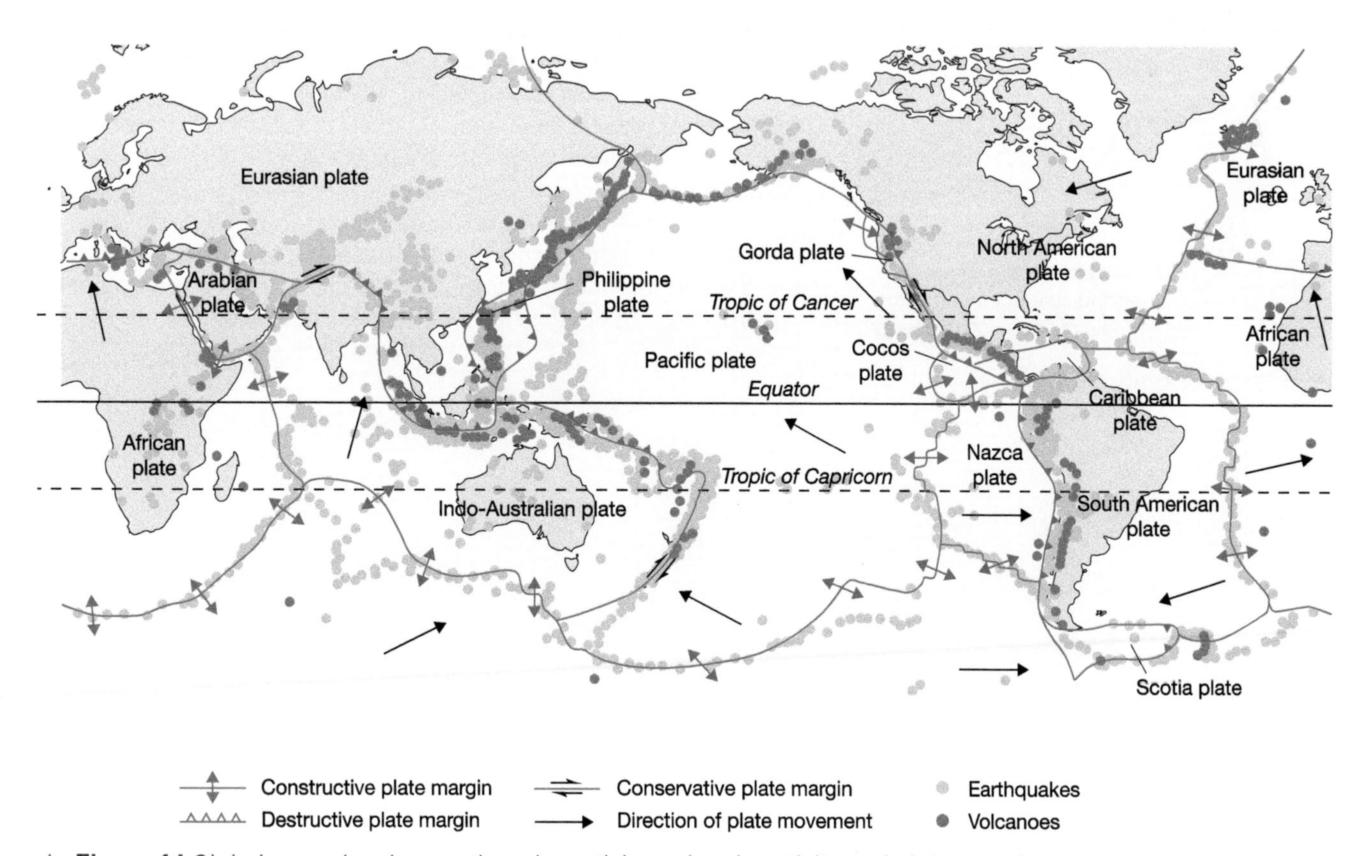

Figure 14 Global map showing earthquake activity, volcanic activity and plate margins.

With more earthquakes being detected each year it might seem that earthquake activity is increasing. However, this increase is simply due to more accurate seismographs being installed and a rise in the number of seismographs feeding into the system.

What are the focus and epicentre of an earthquake?

The main features of an earthquake are shown in Figure 15. The place where an earthquake starts is termed its focus. Shock waves spread from this point, like ripples on a pool after a stone is dropped in. The most deaths and maximum destruction are normally seen right at the epicentre, the point directly above the focus, where shock waves are first felt on the surface. The amount of damage decreases as distance increases from the epicentre. This is described as a negative relationship.

Earthquake location	Date	Strength on Richter scale	Type of country	Deaths
Gujarat, India	2001	7.7	LEDC	20,000
Hokkaido, Japan	2004	7.0	MEDC	0
Northern Sumatra	2004	9.2	LEDC	187,000
Solomon Islands	2010	7.1	LEDC	0
Rat Island, Alaska	2014	7.9	MEDC	0
Nepal	2015	7.3	LEDC	218

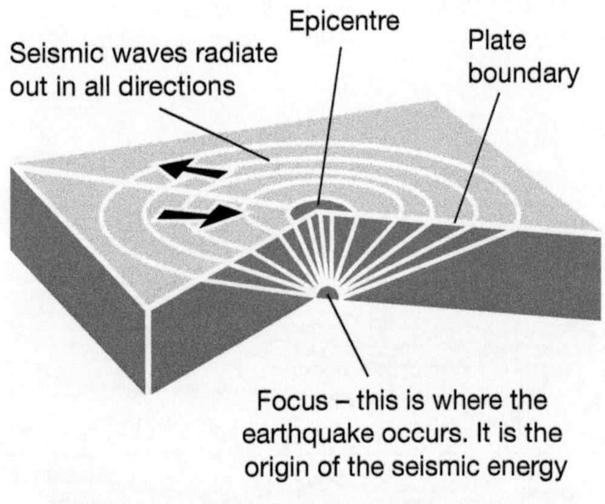

Figure 15 Features of an earthquake.

Activities

1. Make a copy of Figure 15, leaving out a few of the key terms. In the next lesson complete the diagram and see if you can remember it correctly. Check it carefully!
2. Suggest why we don't hear about earthquakes in the news every day, even though lots are happening.
3. Compare and contrast the earthquakes shown in the table above. Tip: make sure you quote figures from the table and use comparison terms like 'stronger', 'less' and 'higher'.
4. Explain what other information we might need to help explain the differing death rates for these earthquakes.

Measuring earthquakes and explaining their consequences

By the end of this section you will be able to:

- explain how earthquakes are measured
- describe and explain how liquefaction and tsunamis happen.

How are earthquakes measured?

Earthquakes are a natural hazard. Major earthquakes can release the same amount of energy as a large nuclear explosion. The destructive power of an earthquake is measured by the amount of energy it releases. Earthquakes are recorded using seismographs.

The paper tracing of the seismic waves (seismogram) is then converted to a level on the Richter scale. The Richter scale generally ranges between 0–10, with each point on the scale representing an earthquake 10 times the magnitude and which releases 30 times more energy than the point before. Figure 17 shows the whole Richter scale.

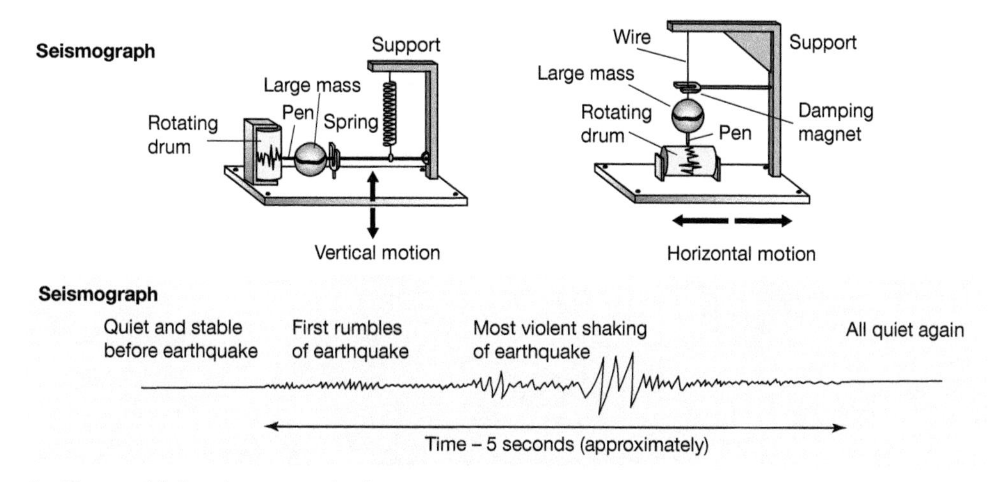

Figure 16 A seismograph diagram.

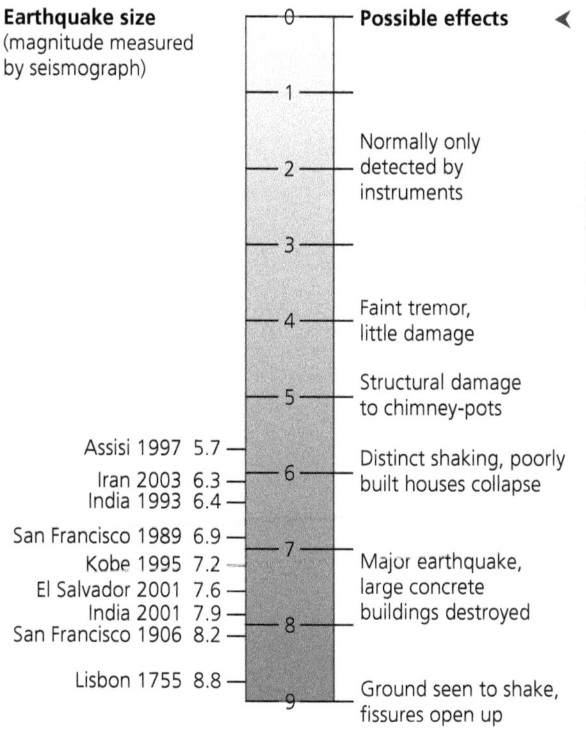

Figure 17 The Richter scale measures the size of the seismic waves during an earthquake.

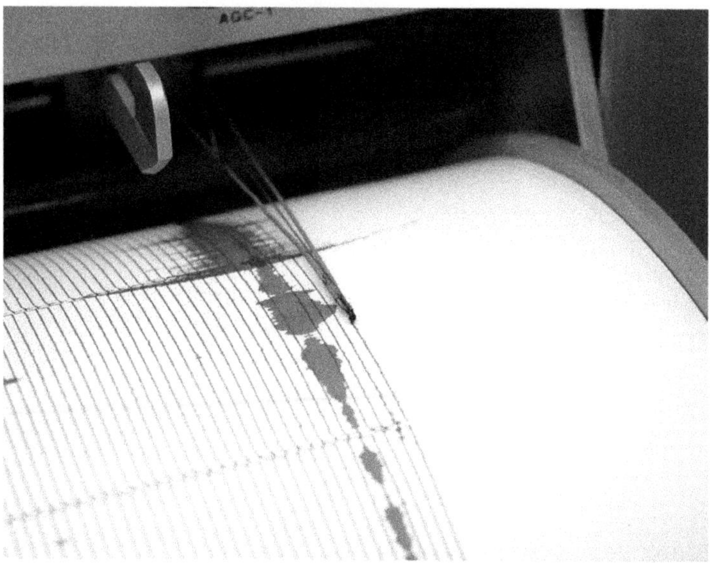

Figure 18 Tracings on the drum of a seismogram show when seismic waves have occurred.

What are the physical consequences of earthquakes?

Earthquakes can have physical impacts on the place where they happen. Two of the main ones are liquefaction and tsunamis.

Liquefaction happens when an earthquake shakes wet soil. The water rises to the surface and turns soil into liquid mud. Buildings can sink into this mud.

A tsunami is a large wave of seawater that can travel for thousands of miles and is triggered by an underwater earthquake. Every time you slide around in the bath and make the water create waves that slap over the edge of the bath you have created your own mini tsunami.

The case study of the Indian Ocean Earthquake (2004) will help you to understand more about the consequences of earthquakes.

Tip 

Try recreating liquefaction by putting wet sand or soil into a plastic container, then placing a solid object like a stone on top. Shake the container to bring the water to the surface and see the stone get swallowed up as the ground liquefies.

▲ **Figure 19** As this road experienced an earthquake and liquefied, an abandoned car sank into the ground.

▲ **Figure 20** This image of the impact of a tsunami was taken in Kesennuma Port in 2011.

Activities

1. Briefly describe how a seismograph measures earthquakes.
2. Describe liquefaction and explain why it happens.
3. Suggest how it is possible that in recent times some earthquakes have measured more than 9 on the Richter scale. You might need to do research on the internet to help your answer.

CASE STUDY

Indian Ocean earthquake, 2004

An earthquake event in an LEDC

Tip

Learn the names of the plates that meet and along whose margin the earthquake happened.

- On 26 December 2004, as many people went about their normal business and holidaymakers started to make their way down to the sun-soaked beaches, a magnitude 9.2 earthquake occurred off the west coast of Sumatra, Indonesia, in the Indian Ocean. Some of the people that the earthquake's effects later killed would not have even known the earthquake had happened. It was the second largest earthquake ever recorded on a seismograph and had the longest-recorded duration, lasting almost 10 minutes.

The plates involved

- There is a major fault line where the Australian plate meets the smaller Sunda plate at the Sunda trench. It is part of a subduction zone. A 15 m slippage along this fault line happened in two stages and led to this prolonged earthquake. As well as the plates moving sideways, the ocean floor rose by several metres and caused the devastating tsunami that marked this infamous earthquake. This raising of the sea floor significantly reduced the capacity of the Indian Ocean, producing a permanent rise in the global sea level by an estimated 0.1 millimetres.

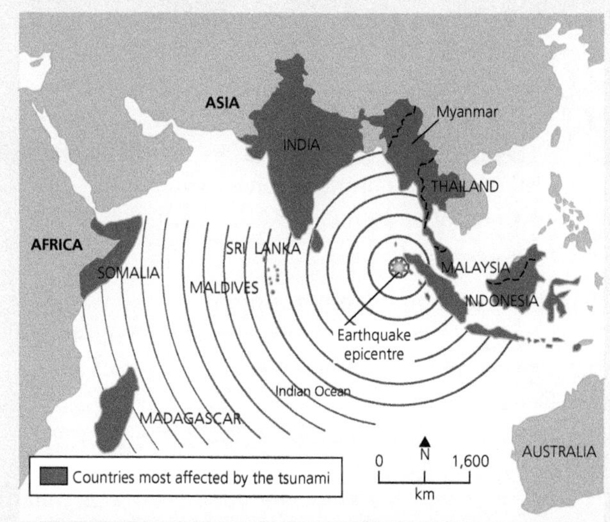

Figure 21 Location map showing the epicentre of the 2004 Indian Ocean earthquake.

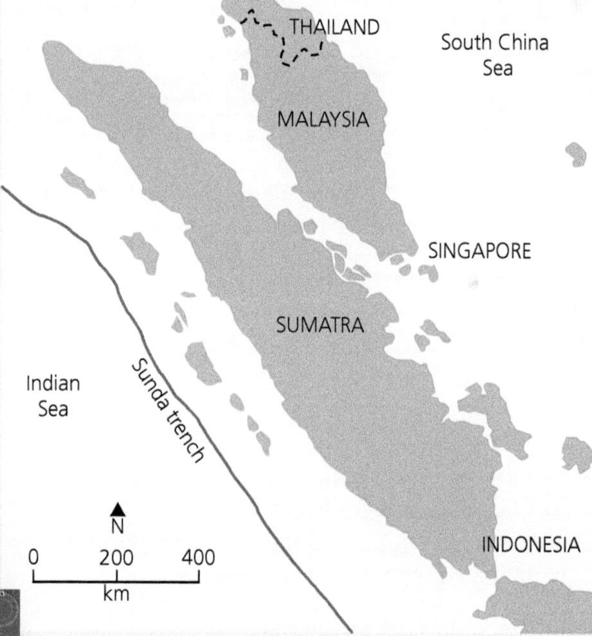

Figure 23 A map of the Sunda trench region.

Figure 22 Satellite image of the Sunda trench region.

Short- and long-term impacts on people and environment

Figure 24 Emergency aid was needed to rescue people of all ages. This photograph shows South Korean firefighters who came to help in the search of ruins of hotel buildings in Khao Lak beach resort on Thailand's western coast.

Short-term impact on people

- Sixty-six per cent of the fishing fleet of Sri Lanka was destroyed. This had important economic implications as fishing provided direct employment to a quarter of a million people on the island.
- Conservative estimates record that over 125,000 people were injured.
- Just over 1.1 million people were temporarily displaced due to coastal devastation.

Short-term impact on the environment

- A tsunami wave reaching 30 m high radiated out from the epicentre, affecting countries on all sides of the Indian Ocean.
- The whole Earth vibrated by 1 cm due to the energy released by this slippage. The energy released on the Earth's surface was the equivalent of 1,502 times that of the atomic bomb exploded in Hiroshima.
- Aftershocks triggered by the main earthquake continued to shake the region for another three to four months.

Long-term impact on people

- The confirmed death toll was just under 187,000. About one-third of this total were children.
- In the Maldives, 17 coral atoll islands had their freshwater supply contaminated by seawater after being overwhelmed by waves. This has rendered them uninhabitable for decades.
- There has been widespread mental trauma as it is a traditional Islamic belief that relatives of the family must bury the body of the deceased, but in many cases no body was retrieved for such a burial.
- One positive result was that the tragedy became the reason why a rebel group declared a ceasefire with the Indonesian government.

Long-term impact on the environment

- The raising of the seabed reduced the capacity of the Indian Ocean and raised the global sea level by 0.1 mm.
- The coastal ecosystems of the areas affected by the tsunami have been severely damaged, including mangroves and coral reefs.
- The massive release of energy is expected to shorten the length of a day by 2.68 microseconds because of a change made to the shape of the Earth.

How prepared was Indonesia for such an earthquake?

What was the management response?

- Although the area affected by this earthquake is within a zone of activity which is used to minor quakes and volcanic eruptions, it is made up of poor LEDCs and they lack the resources to have the scale or quality of response that MEDC areas like Japan can afford.

Before the earthquake: prediction and precautions

- Prior to this earthquake there was no early warning system in place to record such underwater quakes within the Indian Ocean. One island called Simeulue did evacuate its coastal areas. People felt the tremor and fled to inland hills before the tsunami struck. In northern Phuket, an island west of Thailand, a young geography student called Tilly Smith recognised the warning signs of an approaching tsunami wave. She and her parents warned others and the beach was safely evacuated. Most places had no such warnings, so thousands of people were on the beaches enjoying a holiday or preparing to begin a day of fishing. Watch her story on http://movingimages.wordpress.com/2007/10/10/geography-lesson-that-saved-many-lives-the-story-of-tilly-smith.

Immediate and long-term action after the earthquake

- Immediately after the earthquake the world pulled together to provide aid and expertise to the stricken areas. Over US$7 billion dollars were donated from national governments and non-governmental organisations (NGOs).
- Many countries such as Sri Lanka, Indonesia and the Maldives declared a state of emergency to allow strict laws to be implemented to keep order and help with humanitarian aid distribution.
- A review of the poor earthquake and tsunami warning system around the Indian Ocean took place and in June 2006, 25 new seismographic stations relaying information to national tsunami information centres became operational.

▲ **Figure 25** One task that faced countries was clearing up the beaches and burying the dead to reduce the threat of disease. Elephants were used to help.

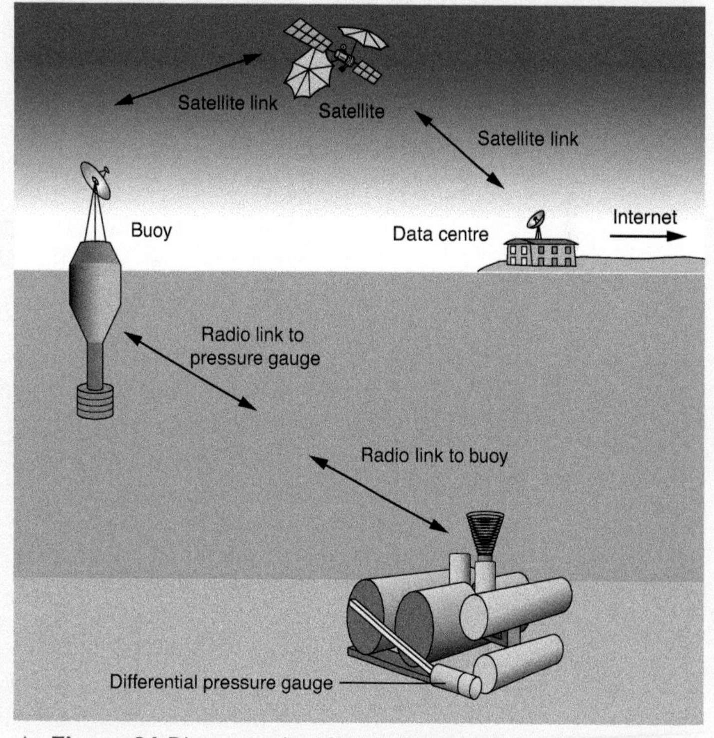

▲ **Figure 26** Diagram showing one positive outcome of this earthquake: a tsunami warning system for Indonesia.

How prepared is Indonesia for a similar earthquake today?

➤ **Figure 27** A memorial park has been built around a ship that had been washed inland over 5 miles by the tsunami.

- By 2014, ten years after the earthquake and tsunami, the city of Banda Aceh has been rebuilt.
- There is now a park and memorial around the site where a 2,600 ton (2,400 tonne) ship had been washed over 5 miles (8 km) inland by the tsunami. The population has risen back to 250,000, which is almost the same as in 2004. There are new highways and vibrant night life.
- However, as a whole, the country's location on the border of a number of dangerous fault lines between tectonic plates means that another large earthquake is inevitable. One of the most notable is the Sunda a megathrust fault line, which runs parallel to the islands of Sumatra and Java. This is the origin of the devastating 2004 quake and even with the new tsunami warning system in place, an instantaneous alert would only give coastal residents 30 minutes to evacuate. This is not enough time to reach higher ground.
- Banda Aceh now has three or four evacuation centres with open ground floors to allow a tsunami wave to pass through, and there is a network of sirens to warn people that a tsunami is imminent.
- Their preparedness was tested on 11 April 2012 when a magnitude 8.6 earthquake struck Banda Aceh. The National Tsunami Warning Centre issued a tsunami alert within five minutes of the first tremor. The on-the-ground response was worrying.
  - 'The conditions were totally chaotic', said Syarifah Marlina Al Mazhir, a lifelong resident of Banda Aceh who worked for the Red Cross during the 2004 tsunami. 'Instead of evacuating to safe areas, people were going home or picking up the kids at school, which created traffic jams.'
  - Even worse, she says, the staff responsible for operating the tsunami sirens fled, and the city's three multi-storey tsunami shelters were locked.
- It seems that the government needs to establish better evacuation routes and education programmes to ensure citizens know what to do. Given the cost of such measures, it seems unlikely that they will be implemented and scientists have pinpointed Padang (a coastal city of 1 million residents) as the next area of Indonesia that is overdue a large earthquake event.
- Thankfully, this plate movement was a horizontal slippage, so did not generate a tsunami and there were only four casualties in Banda Aceh.

As fast as a commercial jet

- Where the ocean is deep, tsunamis can travel unnoticed on the surface at speeds of up to 500 mph (800 km/h), meaning they can cross an ocean in a day or less. Scientists are able to calculate arrival times of tsunamis in different parts of the world based on their knowledge of water depths, distances and when the event that generated them occurred.
- A tsunami may be less than 30 cm in height on the surface of the open ocean, which is why they are not noticed by sailors. But the powerful shock wave of energy travels rapidly through the ocean, as fast as a commercial jet. Once a tsunami reaches shallow water near the coast, it is slowed down. The top of the wave moves faster than the bottom, causing the sea to rise dramatically. The wave of the 2004 tsunami was 30 m high.

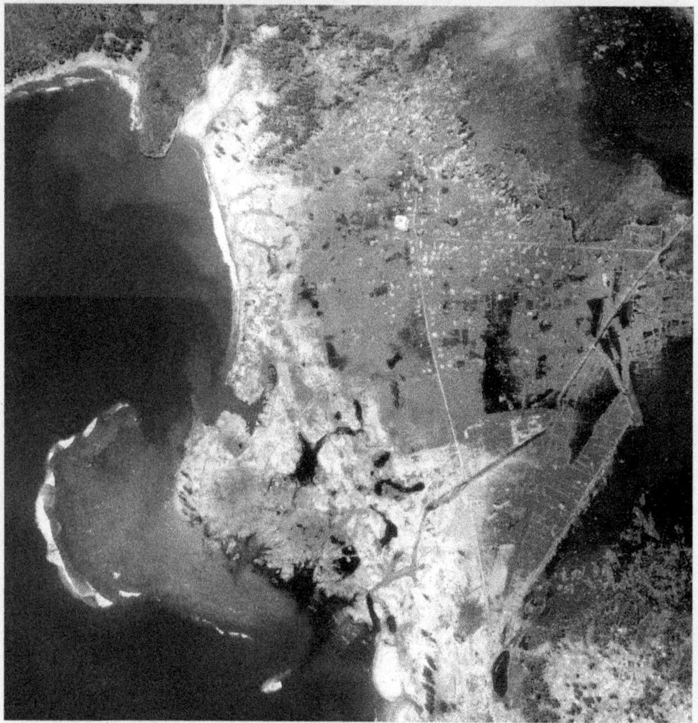

➤ **Figure 28** Stages of mapping the 2004 tsunami with the use of GIS to show the impact of the tsunami on one small area of coastline. The town of Lhoknga, on the west coast of Sumatra, Indonesia, was completely destroyed by the tsunami, as you can see in the bottom photo, with the exception of the mosque (white circular feature) in the city's centre.

Activities

1. Describe a tsunami and explain why they happen.
2. Imagine you were there when the tsunami hit in 2004. Write a poem or a journal entry to describe what you saw and how you felt.
3. For a named earthquake you have studied, explain why it happened and evaluate the measures taken before it to prevent deaths and damage.

By the end of this section you will be able to:

- identify shield, composite and supervolcanoes, and describe their characteristics.

Tip

Ensure you can compare the three types of volcanoes as well as simply describe them.

The characteristics of volcanoes

What are the characteristics of shield, composite and supervolcanoes?

Volcanoes are mountains, often cone-shaped, that are formed by surface eruptions of magma from inside the Earth. During eruptions, lava, ash, rock and gases may be ejected from the volcano. Areas that experience volcanic activity form three main belts:

- around the edge of the Pacific Ocean – known as the 'ring of fire' because there are so many active volcanoes
- through the Mediterranean Sea, and down the east coast of Africa
- down the middle of the Atlantic Ocean.

Volcanoes are also found in isolated clusters, such as the Hawaiian Islands in the middle of the Pacific Ocean, and Réunion in the Indian Ocean.

Shield volcanoes

Shield volcanoes are large, wide, cone-shaped volcanoes that have gentle slopes as they are made from runny (basic) lava flows that have hardened on top of each other.

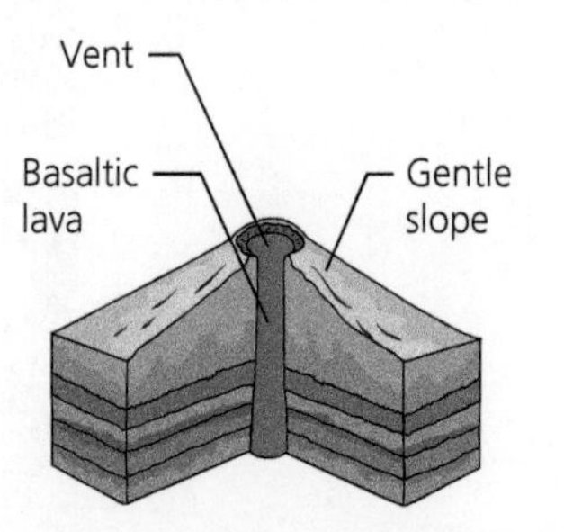

Figure 29 A gently sloping shield volcano made from free-flowing basaltic lava.

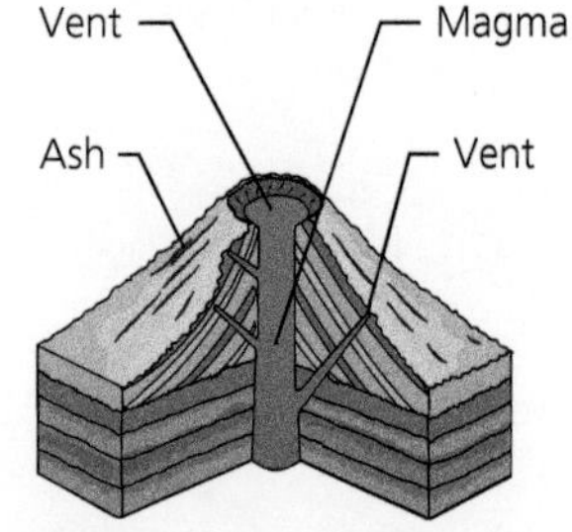

Figure 30 A tall composite volcano made from alternate layers of ash and lava.

Composite volcanoes

Composite volcanoes are classic mountain-shaped volcanoes where the cone has been made up from hardened layers of ash and lava created after each eruption. They occur at destructive plate boundaries and have common features such as a high cone, main and secondary vents.

Supervolcanoes

Supervolcanoes have a rather different structure than other volcanoes (see Figure 31). Rather than a cone shape, they create wide depressions called calderas, with a ridge of high land encircling them. A caldera forms when a volcano erupts so violently that it collapses in on itself or over a hot spot or volcano that has caved in on itself, covering the magma chamber under layers of rocks. The magma and pressure build up over time, ending in a violent eruption which can disrupt the whole world. Campi Flegrei near Naples is a dangerous supervolcano that in December 2016 was described as 'awakening'.

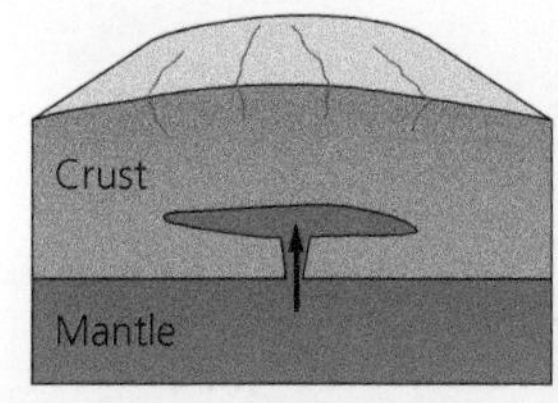

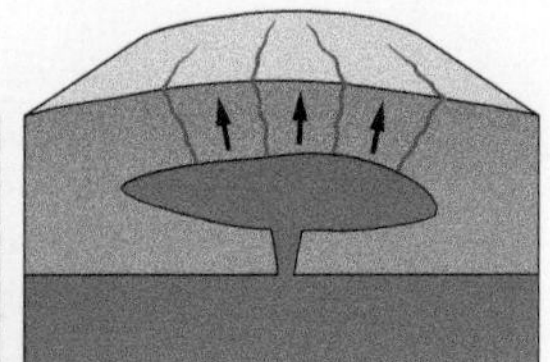

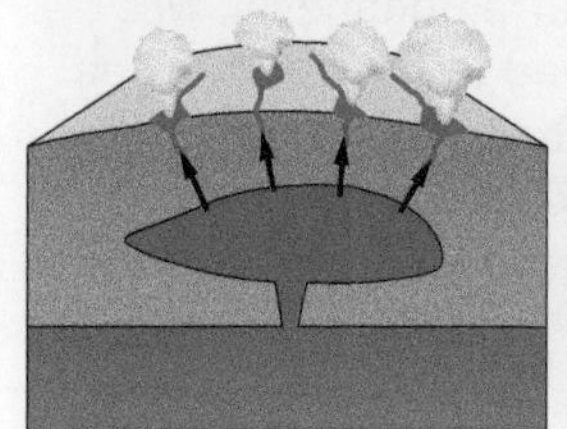

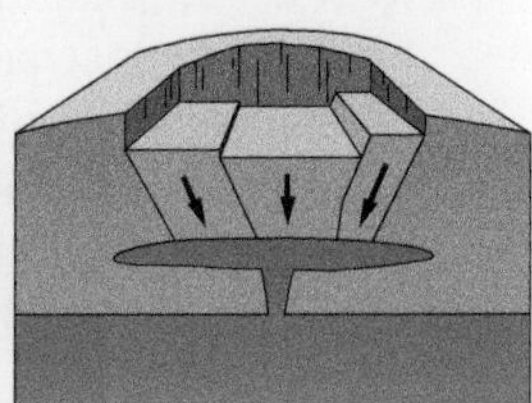

Figure 31 The formation of a supervolcano.

Activities

1. Where is the 'ring of fire' and why did it get this name?
2. Compare the characteristics of a shield volcano to a composite volcano.
3. What is a 'caldera' and how is it related to the existence of supervolcanoes?

The potential global impacts of a supervolcano eruption

By the end of this section you will be able to:

- discuss possible global impacts on people following a supervolcano eruption
- discuss the possible global impacts on the environment of a supervolcano eruption.

What are the potential global impacts of a supervolcano eruption?

To be classed as a supervolcanic eruption, at least 1,000 km^3 of material must be erupted during the explosion. Compare this to the largest eruption of recent times, Pinatubo in the Phillipines (1991), when only 5 km^3 of material was erupted.

Yellowstone National Park, Wyoming, USA

Around 4 million people each year visit Yellowstone National Park to see the amazing wildlife and view the geothermal wonders of geysers and bubbling mud pits. The world's first national park, visitors may not realise that this whole region is actually a supervolcano.

Tip 

Learn some facts and figures about the potential impacts of a supervolcanic eruption from the information on Yellowstone. Use these to add depth to longer responses about these types of eruption.

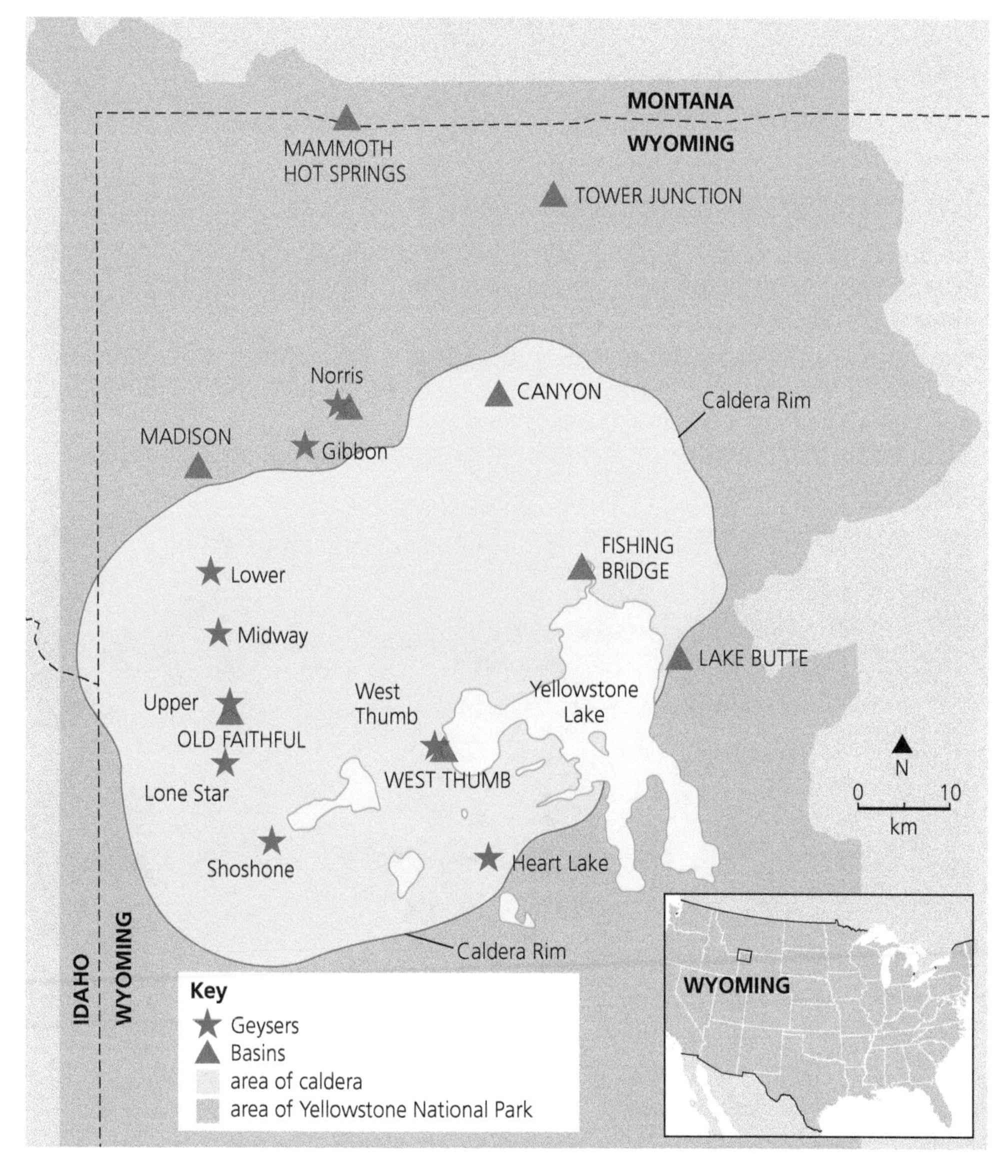

▲ **Figure 32** Map of Yellowstone National Park, Wyoming, USA.

Beneath Yellowstone is a shallow and very large magma chamber (see Figure 33). New technology has allowed geologists to estimate the chamber to be 80 km long and 20 km wide. This is larger than anyone had previously thought. Over recent years there have been earthquake swarms across the caldera and vertical ground movements of up to 8 cm. However, such movements have been observed for years without preceding an eruption. No volcanic eruption has occurred at Yellowstone for 70,000 years.

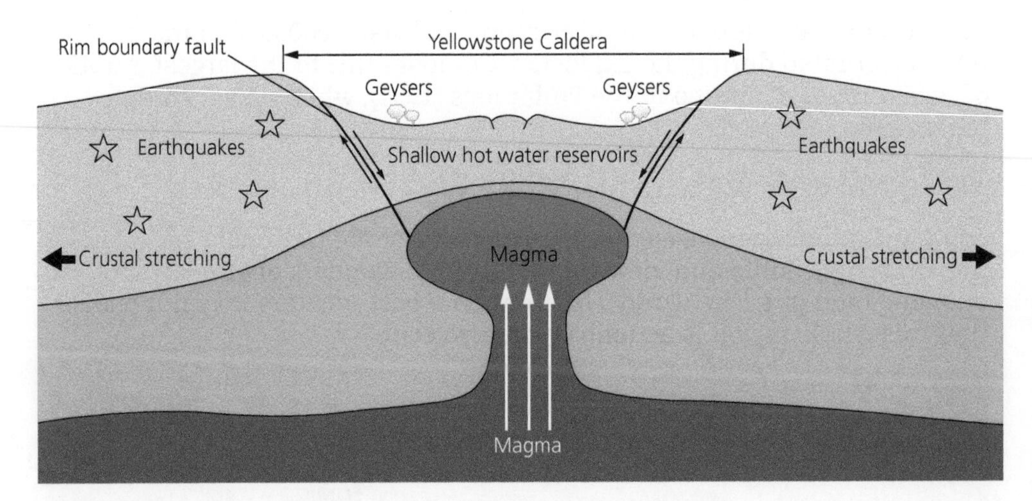

Figure 33 A cross-section of Yellowstone volcano.

The potential impacts on people

If a super eruption did happen at Yellowstone, the USA would be the worst affected. The greatest danger would be within 1,000 km of the blast, where almost everyone would be killed.

Ash is one of the most dangerous substances that erupts from volcanoes. As supervolcanoes are highly explosive, most of the lava will be blasted into the air to create such ash if Yellowstone did erupt. Volcanic ash particles are not smooth; under the microscope it can be seen they are jagged (see Figure 34).

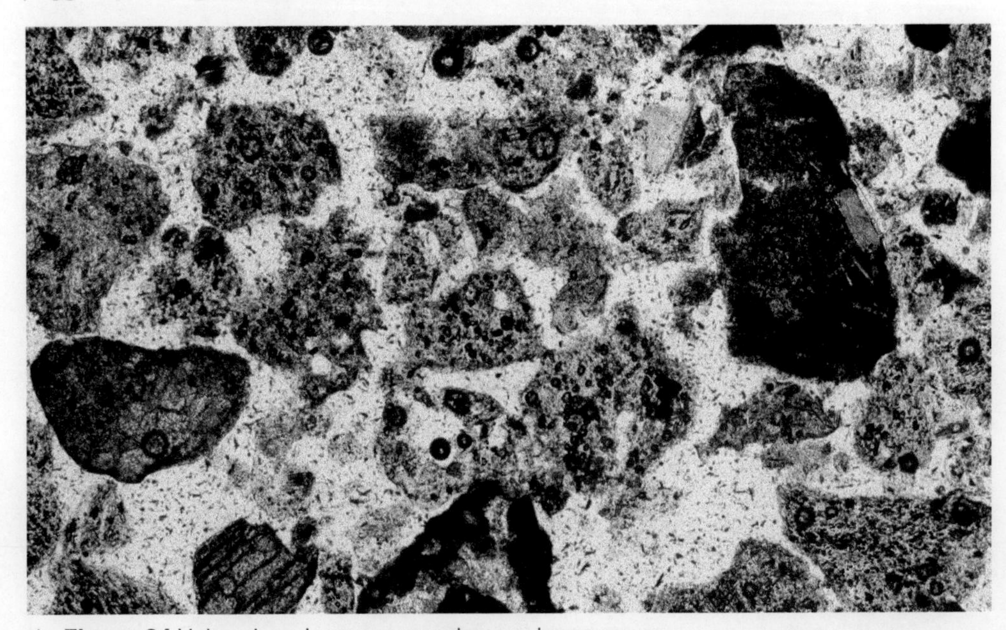

Figure 34 Volcanic ash as seen under a microscope.

An estimated 90,000 people might die in total from inhaling ash, which forms a cement-like mixture in human lungs. Even the east coast could be immobilised by just 1 cm of ash.

Many buildings would be destroyed near Yellowstone as it only takes 30 cm of dry ash to cause a roof to collapse. Water supplies would become undrinkable and the ash would quickly clog up air filters on all vehicles from cars to aeroplanes, so transport across the USA would be severely disrupted for weeks. Of course, air travel to and from the USA would also be badly affected, causing major disruption in other countries and to businesses.

The potential impacts on the environment

A Yellowstone eruption could inject 2,000 million tonnes of sulphur high above the Earth's surface. Once there it would take two to three weeks for the resulting sulphuric acid aerosols to cloak the globe. Once formed, these aerosols reflect sunlight, reducing the amount of energy reaching the lower atmosphere and the Earth's surface, cooling them.

Global annual average temperatures would drop by up to 10 degrees Celsius, according to computer predictions. And the northern hemisphere could cool by up to 12 degrees Celsius. Experts say colder temperatures could last six to ten years. Such a dramatic drop in temperature could mean crop failures and a return to the 'Little ice age' conditions seen in Europe during the late 1600s.

Locally, the 67 species of mammals that live in Yellowstone national park, including bison and wolves, would most certainly die, thus disrupting the local ecosystem for decades. Indeed, scientists that have studied evidence from previous eruptions estimate it is at least ten years before any vegetation becomes re-established in the areas close to a super eruption.

Activities

1. Explain the meaning of the term 'supervolcano'.
2. Research the global location of supervolcanoes and write out the names and locations of three such supervolcanoes.
3. Write out three differences between a volcano and a supervolcano.
4. Describe the location and size of the Yellowstone supervolcano.
5. Imagine Yellowstone has erupted. Write a report to be read by a local news reporter on the radio describing the impacts for the area close to the eruption and later for the UK. Your report will need to be clear in its description as there are no images in a radio broadcast.

Sample examination questions

1 Complete the following sentences by underlining the correct word. [4]

(i) Plates are part of the earth's **crust / core**.

(ii) Plates move on top of the **crust / mantle**.

(iii) Heat from the **crust / core** sets up convection currents in the mantle.

(iv) Convection currents cause the **movement / liquefaction** of plates.

2 Describe one characteristic of the Earth's inner core. [2]

3 Complete Table 1 by filling in the empty boxes about some common rocks. One has been completed for you. [4]

Name of rock	Type of rock	Common colour of rock
sandstone	sedimentary	Yellow/red or light brown
slate		
basalt		

Table 1 Common rocks.

4 Explain how sedimentary rocks such as sandstone were formed. [3]

5 Describe two impacts caused by an earthquake you have studied. [6]

6 Explain why volcanoes occur at destructive plate margins. [3]

7 Explain how fold mountains are formed at a collision plate boundary. [3]

8 Complete Table 2 by stating whether each statement about earthquakes is true or false. [2]

Statement	True / false
The focus of an earthquake is above the epicentre.	
Most earthquake damage occurs close to its epicentre	

Table 2 Earthquake facts.

9 Study Figure 1 below which shows the plate margin where the Nazca Plate meets the South American plate. Use it to help you answer the following questions.

(i) In the box below, underline the type of plate margin shown in Figure 1. [1]

Constructive	Destructive	Collision	Conservative

(ii) Name the landforms found at points A and B. [2]

____

____

(iii) Explain why earthquakes are common at plate margins such as this. [3]

____

____

____

____

10 Describe the global distribution of volcanoes shown in Figure 2. [4]

____

____

____

____

____

____

____

Tip

Do not simply list or name some volcanoes. Use terms like 'linear pattern' and give directions, with general location, using terms like 'west coast of South America'.

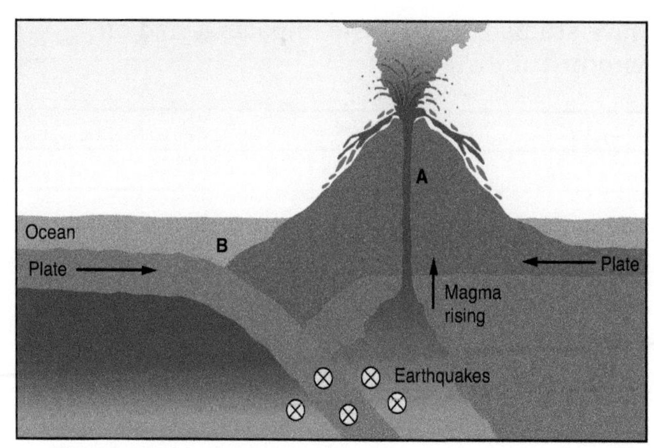

Figure 1 Plate margin where the Nazca plate meets the South American plate.

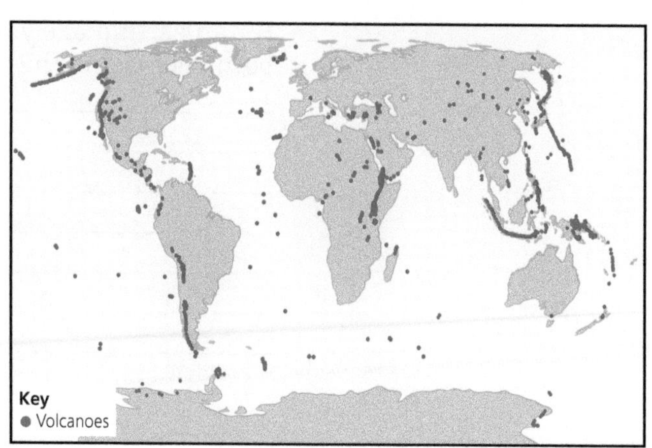

Figure 2 The global distribution of volcanoes.

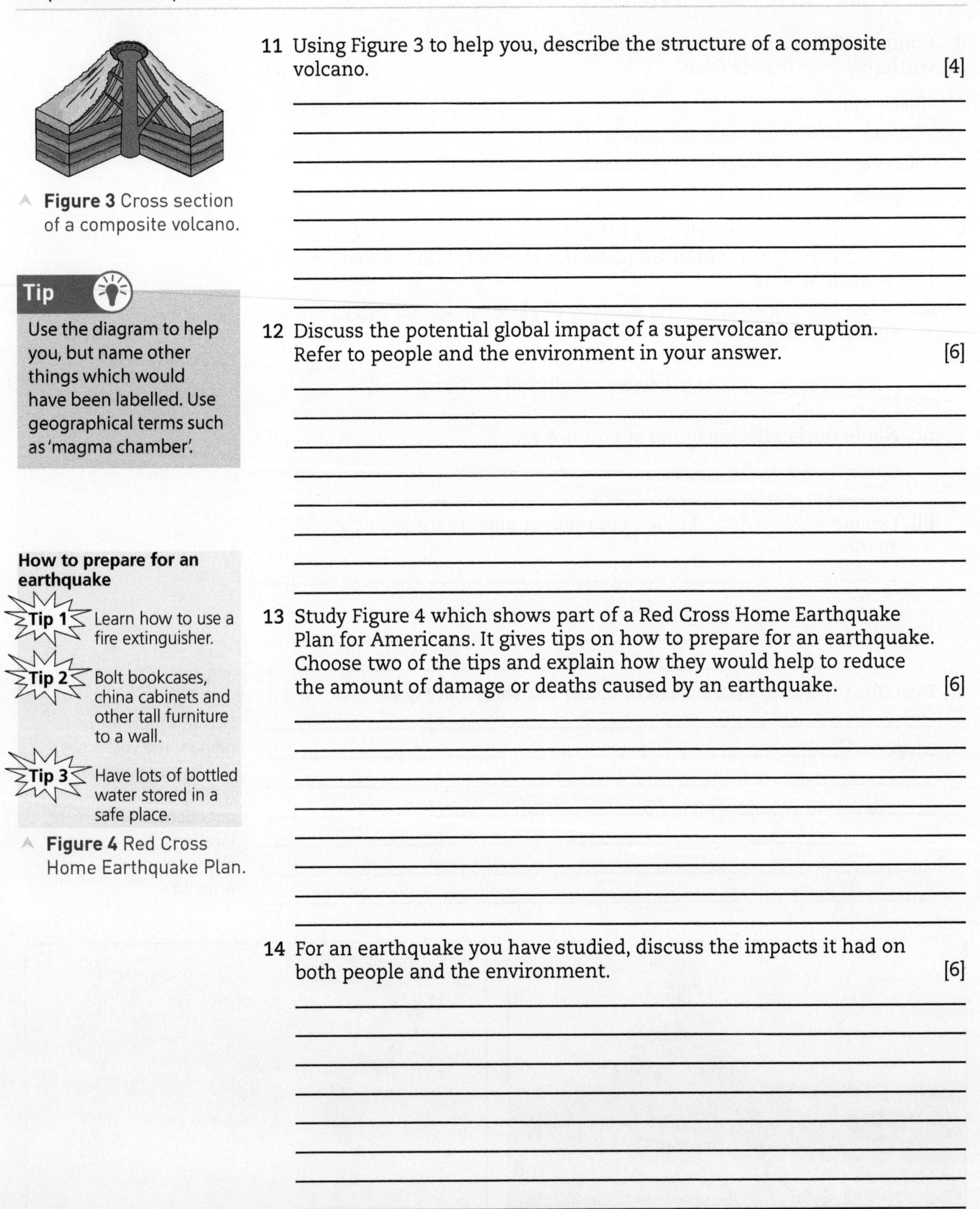

Figure 3 Cross section of a composite volcano.

Tip

Use the diagram to help you, but name other things which would have been labelled. Use geographical terms such as 'magma chamber'.

11 Using Figure 3 to help you, describe the structure of a composite volcano. [4]

12 Discuss the potential global impact of a supervolcano eruption. Refer to people and the environment in your answer. [6]

How to prepare for an earthquake

Tip 1 Learn how to use a fire extinguisher.

Tip 2 Bolt bookcases, china cabinets and other tall furniture to a wall.

Tip 3 Have lots of bottled water stored in a safe place.

Figure 4 Red Cross Home Earthquake Plan.

13 Study Figure 4 which shows part of a Red Cross Home Earthquake Plan for Americans. It gives tips on how to prepare for an earthquake. Choose two of the tips and explain how they would help to reduce the amount of damage or deaths caused by an earthquake. [6]

14 For an earthquake you have studied, discuss the impacts it had on both people and the environment. [6]

UNIT 2

LIVING IN OUR WORLD

THEME A: Population and Migration

International border between Mexico and the USA.

What are the barriers to international movement?

By the end of this section you will be able to:

- define demographic key terms.

Defining crude birth rate, crude death rate and natural change

Demographers (people who study populations) use some key terms which you will need to know too for your study of geography.

What does crude birth rate mean?

Population growth occurs when birth rates are greater than death rates. Measuring birth rates involves more than just counting the total number of babies born in a year. The populations of Belize and Belgium (see Figure 1) are good examples to demonstrate this.

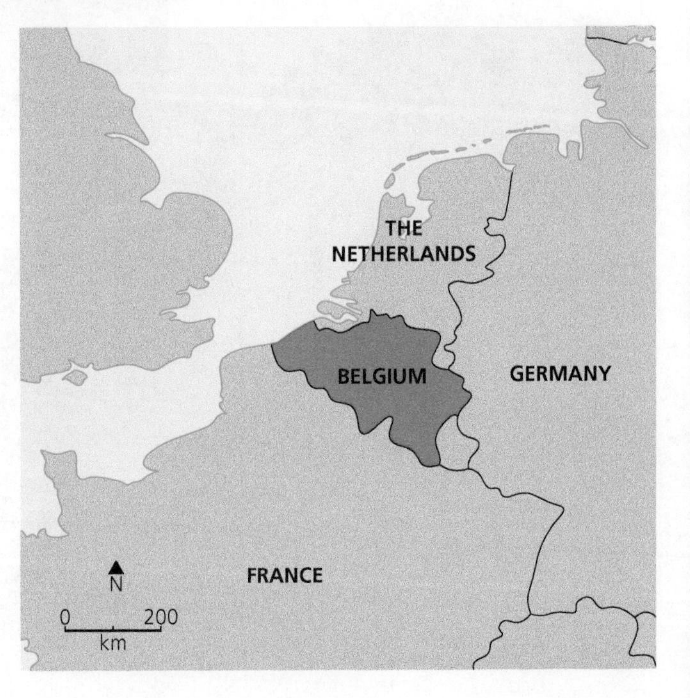

Figure 1 Belize and Belgium.

Comparisons (2016 est. data)	Belize	Belgium
Location	Central America	Europe
Level of development	LEDC	MEDC
Births recorded	8,452	131,403
Deaths recorded	2,028	111,109
Total population	371,423	11,413,049

Source: CIA World Factbook and http://countrymeters.info

To compare birth rates in these countries properly, we should take into account the size of their populations. This can be done by working out the number of births per thousand of the population every year. This is called the crude birth rate.

A figure of 40 or more births per thousand people per year is a very high birth rate. Countries with low birth rates have figures of around 10 per thousand per year. Belize's crude birth rate is 22.8 per thousand per year, compared to just 11.5 in Belgium. There may be more babies born in Belgium but, when you take the size of the populations into account, Belize has a much higher birth rate.

What does crude death rate mean?

Similar to the crude birth rate, the crude death rate is a measure of the number of people who die in a country each year, per thousand of the population. Staying with our two example countries, Belgium has a crude death rate of 9.7 per every thousand people while Belize has a crude death rate of 5.5 per thousand. It might look surprising that an LEDC country has a lower death rate than an MEDC country, but you have to remember that many LEDCs have a very young population: around 36 per cent of the population of Belize is under 15, and only 4 per cent are 65 and over. MEDCs generally have an ageing population: Belgium has just 16 per cent of its population under 15, with almost 18 per cent aged 65 and over. As the main causes of death in the world are heart disease, strokes and lung diseases related to smoking, young people are more likely to survive. Countries with a youthful population are likely to have lower crude death rates compared to countries with ageing populations.

What does natural change mean?

The difference between the crude birth rate and the crude death rate is known as natural change. If the crude birth rate is higher than the crude death rate, there will be a natural increase in the population, and if more people die in a country than are replaced by births, there will be a natural decrease. It is like having a bath with one tap and one plug hole (see Figure 2).

Of course, measuring population growth and change is more complicated than this, as people are constantly migrating in and out of countries. You would need another tap (immigration) and another plug hole (emigration) for the model to work. When looking at the population of the whole world, as opposed to individual countries, changes are measured as a result of natural increase and decrease only.

'If the water coming in from the tap is greater than that being lost through the plug hole, the level of the bath **will rise**.'

'If more water is being lost down the plug hole than is being put in from the tap, the level of the bath **will fall**.'

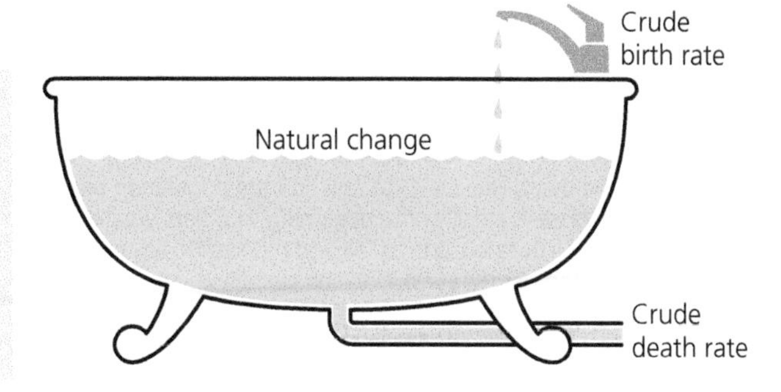

Figure 2 Natural change.

By the end of this section you will:

- know and understand the Demographic Transition Model (DTM)
- understand why the DTM is useful.

The Demographic Transition Model (DTM)

What is the Demographic Transition Model?

A model is a simplified version of something more complicated. Think of a model car, or a model plane. We use models in geography to understand the world better, by simplifying what is happening so that we can understand it more easily. Demographic transition is the change over time (transition) in population (demography). We can look at the Demographic Transition Model (DTM) to see how this happens in most countries.

Populations change for several reasons. People move in and out of countries (migration). Another reason for population changes are increases and decreases in crude birth rates and crude death rates. Demographers measure these in numbers per thousand of the population. Birth rates and death rates may be as high as 40 per thousand per year or below 10 per thousand per year. Figure 1 shows both birth rates and death rates over many years.

Birth and death rates

Look at the death rate line in Figure 3. At the beginning it is around 40 per thousand of the population per year. It then starts to fall, and eventually stabilises at around 10 per thousand per year.

On first glance, 40 births per thousand per year may seem like a low birth rate. Remember that about half of the thousand are male and won't have babies. From the remaining (female) half, most will be too young or too old to become pregnant. This makes a birth rate of 40 per thousand per year actually very high.

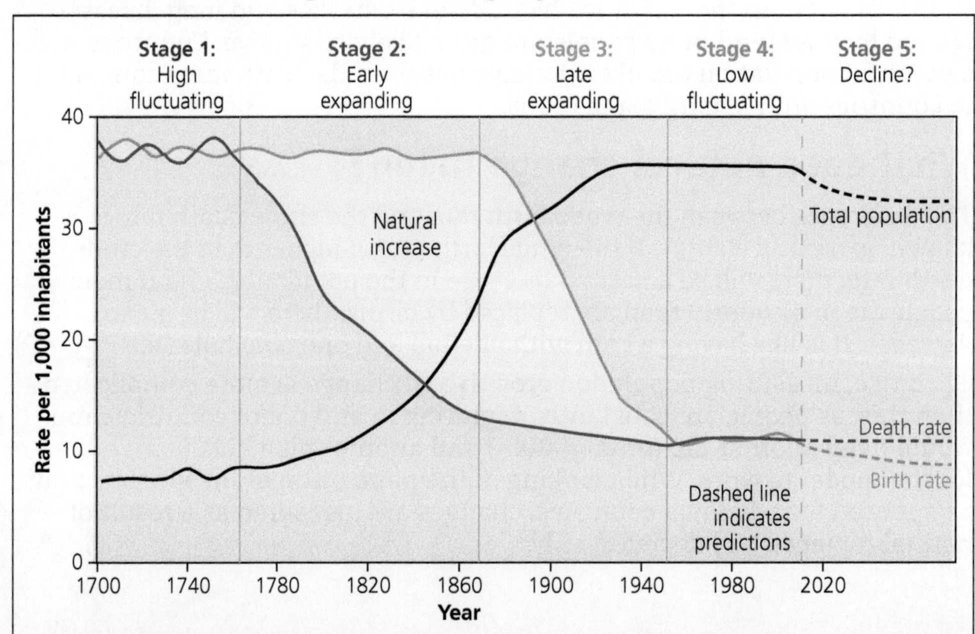

Stage 1: Birth rates and death rates are high so the population does not grow very much, and is small. Very few places in the world are still in this stage, but the UK and Ireland were like this in the early 1700s.

Stage 2: The death rate starts to fall, and the population begins to increase. Notice that the natural increase is small at the beginning, but gets greater over time.

Stage 3: The birth rate starts to fall and natural increase starts to decline. There will be natural increase until the birth rate falls to the same level as the death rate: the population is still growing but the speed of that growth starts to slow down.

Stage 4: The birth rate has fallen to the same as the death rate and the rate of natural increase is very low. The population will stop growing, or grow very slowly. The UK and Ireland are in this stage.

Stage 5: Some demographers believe there is another stage where the birth rate falls below the death rate and stays there. If that were to happen, the population of those countries would decline.

Figure 3 The Demographic Transition Model.

The birth rate line in Figure 3 also decreases from around 40 per thousand per year to less than 10 per thousand per year, although it starts to fall later than the death rate starts to decline.

Reasons for falling death rates	Reasons for falling birth rates
▶ Better health care, for example, by having more hospitals, increased access to medicines, and more knowledge of how diseases are spread. ▶ More doctors, enabling more patients to be treated. ▶ Better hygiene, decreasing the risk of infection. ▶ Common diseases eradicated (for example, smallpox) and others now more easily treated (for example, measles). ▶ Better diet, improving the health of the population so fewer people become ill and others are better able to survive if they do become ill. ▶ Clean drinking water.	▶ Decreasing child mortality: children are more likely to live beyond infancy, so people tend to consider it safe to have fewer children, expecting them to survive. ▶ As the population moves to cities and machinery is used more in agriculture, people do not need to have children to work on the farms. ▶ Better education means that there is more awareness of how families can be planned by using birth control methods. ▶ Increased opportunities for women through education may mean that more will delay starting a family until they are older. ▶ Changing perception of gender roles will accept that females can take responsible jobs and are not confined to housekeeping and child-rearing.

As we have seen, when the birth rate is higher than the death rate, there is natural increase: the population will grow. If the death rate is higher than the birth rate, there is a natural decrease in the population. If birth and death rates are close to each other, with about the same number of people being born as dying, there will be little natural change in the population.

Population change

There is a third line on the graph in Figure 3, which shows total population change. In Stage 1 total population is low, and not growing because the birth rates and death rates cancel each other out. When the death rates fall, the total population starts to grow ever more quickly. It only starts to slow when birth rates begin to fall in Stage 3, but continues to grow until birth rates drop to the same level as death rates. The line is not specific for any country, but shows the general pattern of how populations change over time.

Skills activities 

Here are birth and death rates of a number of countries.

Country	Population	Birth rate per thousand per year	Death rate per thousand per year
Angola	24,227,524	45	14
Finland	5,461,512	11	10
Ireland	4,615,693	14	6
United Kingdom	64,559,135	12	9
Venezuela	30,693,827	20	6

▲ **Source:** The World Bank (2014 figures)

Work out the natural increase (take the death rate away from the birth rate, and divide by 10 to make it a percentage) and, with a calculator, see how quickly the population will grow over five years. Work with a partner or in a group to calculate them all. Angola has been done for you below.

Angola

Natural increase is 45 minus 14, divided by 10: 3.1%

Population in 2014: 24,227,524

So, for each year, divide the population by 100 and multiply by 3.1 to find the increase. Then add that on to the previous year's population. An easier way is to multiple the figure by 103.1.

2015: 24,978,577
2016: 25,752,913
2017: 26,551,253
2018: 27,374,342

Why is the DTM a useful model?

The DTM is the starting point of any study of demographics. It is a simplification of population change within countries, which is far more complicated than the model can allow. However, the demographic processes – those factors which cause the population to change – will be similar in many countries. Using the model means that we can see what might be expected of a county's population over time: it can be used to inform a prediction. The DTM can also give some idea of what the population structure is likely to be as birth and death rates change. Will there be a greater number of older people, or of children?

The DTM was created by looking at what happened to more economically developed countries (MEDCs) in Europe and North America. Some demographers criticise it as Eurocentric and point out that the population processes in other countries, including many less economically developed countries (LEDCs), might be different:

- Some LEDCs seem to have gone through Stage 2 with death rates falling, but have a persistently high birth rate; perhaps because of social or cultural factors. Countries which have not yet moved into Stage 3 include many countries in Africa south of the Sahara desert, Yemen, Afghanistan and Palestine.
- Some LEDCs had a very large population when Stage 2 began, so the impact of the natural increase has been greater. Examples include Nigeria and China.

Additionally:

- China has skipped a stage with its one-child policy bringing birth rate down very rapidly. This policy began to be phased out in 2015, but its effects can still be seen in the population.
- The DTM does not consider how governments can help to control the populations of their countries by their policies.
- The DTM does not consider the impact of migration.

Research activities

1 Find the birth rates and death rates of a country over time, and plot these on a graph. You could try searching in the World Bank site (http://www.worldbank.org/en/country)

2 a How close is your chosen country to the DTM?

b Can you work out when Stage 2 and Stage 3 begin in your chosen country?

c Is the population change in your chosen country more complicated than the model?

d Does the model help you to understand the population change in your chosen country?

Activities

The total fertility rate is the number of children born to each woman over her lifetime, on average. As some have no children, other women will have more than the average figure.

1 What does the map show you about the places where the highest fertility rate is?
2 Name four countries that have the highest fertility rates.
3 Where on the DTM might these countries be? What is your evidence for that?
4 In total, 484,367 people died in England and Wales in 2011. The total population of England and Wales was 56 million in 2011.
   a Calculate the death rate in England and Wales per thousand per year.
   b Suggest which stage of the DTM the UK was in in 2011.

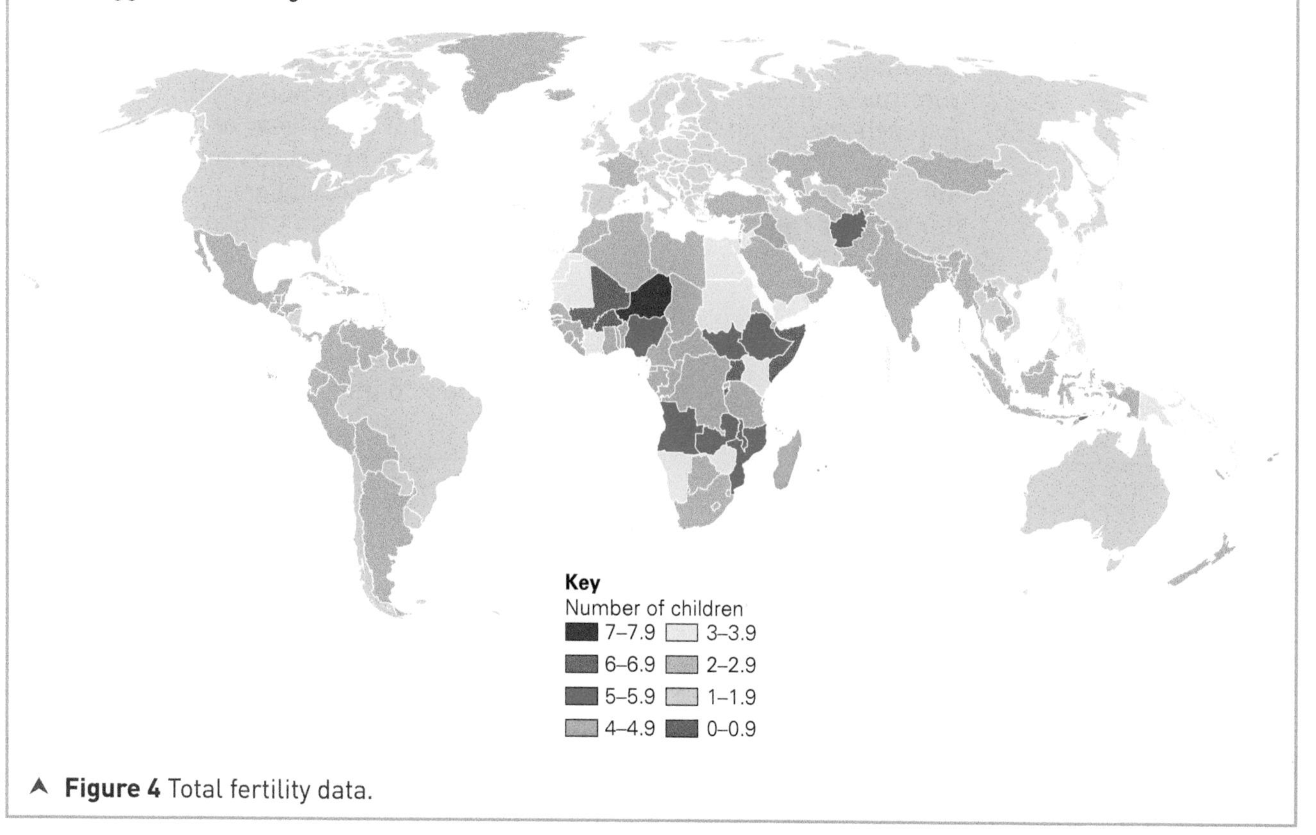

▲ **Figure 4** Total fertility data.

By the end of this section you will:

- know and understand what population structure is
- know how to compare and contrast population structures of different countries.

Population structure

What is population structure?

Population structure is the composition of the people living in a country. One way of showing the composition of the population in a country or region is a population pyramid. This shows two main characteristics: the age and the gender of the people living in an area. This gives the diagram its other name: an age–sex pyramid.

What do population pyramids show?

Population pyramids are divided into two: usually the left-hand side shows the male population and the right-hand side shows the female. The bars that make up the pyramid represent the age of the population, with a bar for each age group. The bars can be drawn for every year or, more often, for five-year groups: ages 0–4 years, 5–9 years, 10–14 and so on. The numbers along the bottom of the pyramid represent how many are in each bar. Sometimes the numbers are the total people in each category, but more often it is the percentage of people in each category. Using percentages is better as it allows pyramids to be compared more easily.

Figure 5 is a population pyramid for the UK from the most recent census in 2011, which displays the information based on the numbers of people in each year group. The labels explain how to interpret what the pyramid is showing.

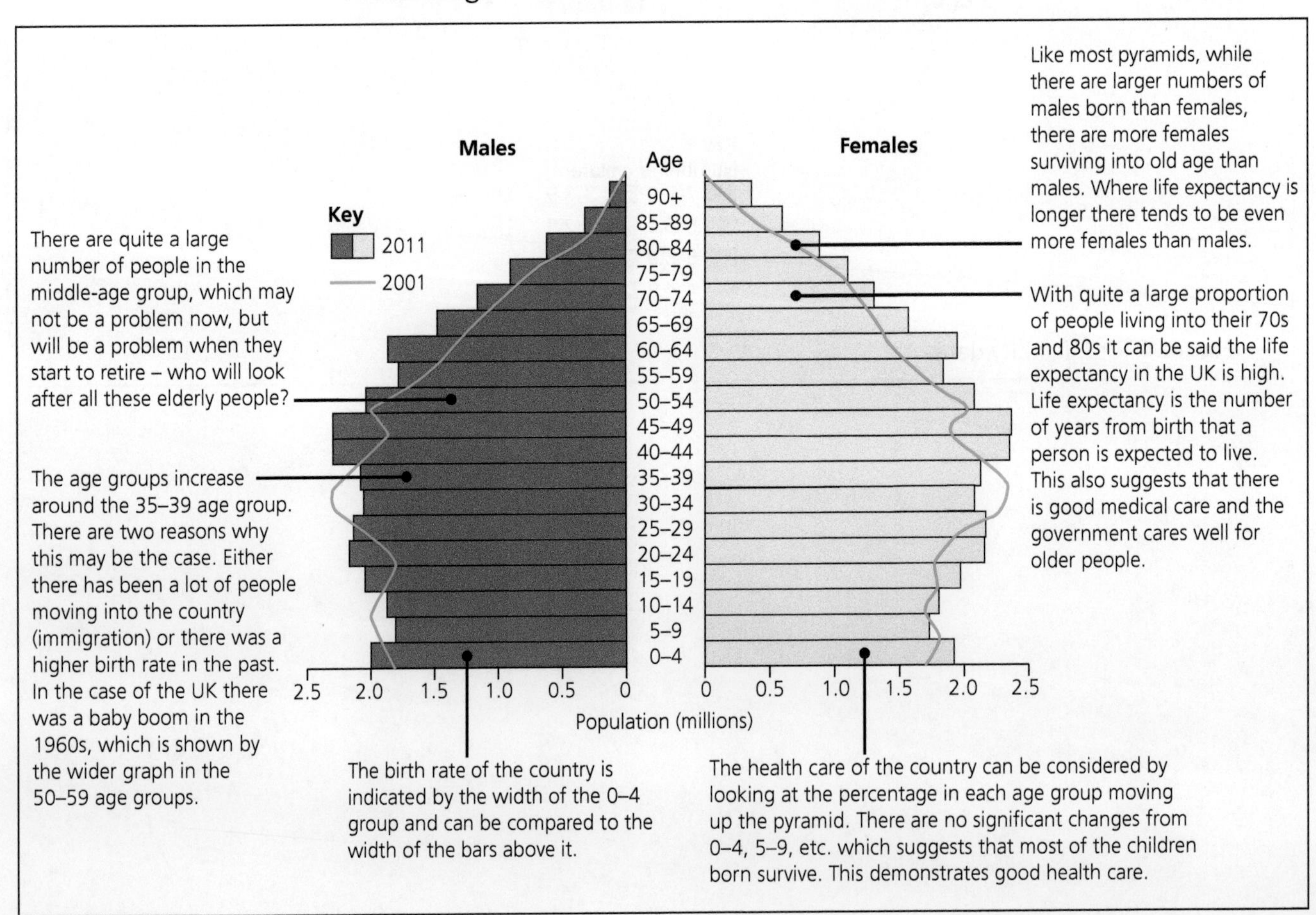

▲ **Figure 5** Population pyramid of the UK for 2011.

How do LEDC population pyramids compare to MEDC pyramids?

In general:

- Less economically developed countries (LEDCs) have a wider base to their pyramids, which indicates a higher birth rate. The rapidly decreasing sides point to this high birth rate with total population rising rapidly. They may also indicate a lot of infant deaths. These pyramids come up to a sharp point, indicating very few elderly people. The population has grown so fast that the proportion of elderly people is tiny. The bulk of the population is young.
- More economically developed countries (MEDCs) have a narrower base, which indicates that the population is growing very slowly or not at all: the birth rates and the death rates are similar, so they more or less cancel each other out. Their sides are not steep like the LEDCs' because their populations are not growing and they have low levels of infant mortality and good health care. These pyramids tend not to taper off until older age groups as there is generally a long life expectancy.

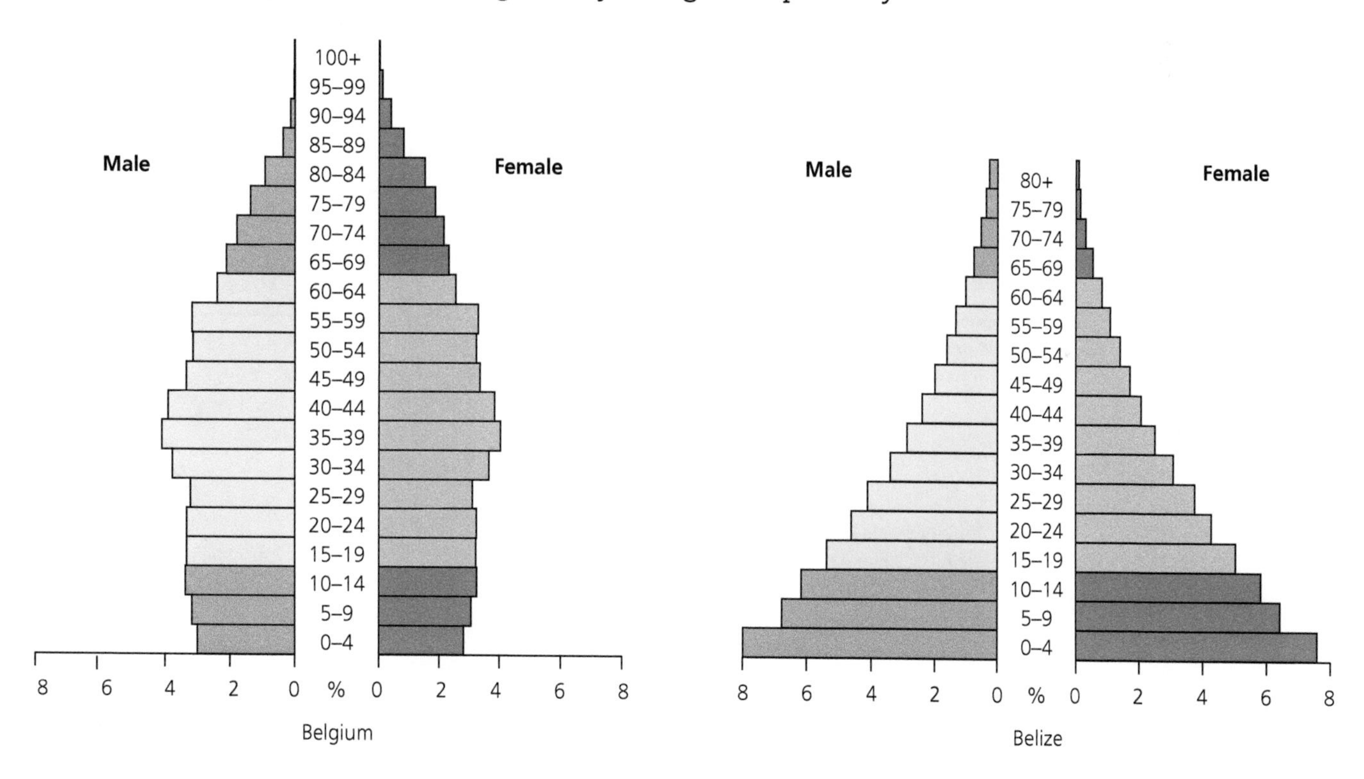

▲ **Figure 6** Population pyramids for Belgium and Belize. Based on 2013 (Belize) and 2016 (Belgium) population figures.

Activities 

1 Find an animated population pyramid for England and Wales (in a search engine put the phrase 'Animated population pyramid gov.uk' and click 'search' – it should be the top link from www.neighbourhood.statistics.gov.uk). The page shows the population pyramid for 1961. Hover over a bar to get the age and gender details of the population.
   a Find the bar for people aged 20 in 1961. Note the small number of people. What was happening all those years before those people were born? The display also shows when they were born. Why was there a lower birth rate during the year they were born?
   b Now find the bars for people aged 12–15 in 1961 (the baby boomers). Look at the dates that they were born. Why did the birth rate go up then?
   c Find people aged around 43 in 1961 – there is a shortage of these people. What was happening in the world when they were born which would have reduced the numbers of people of that age?
2 Watch the animation, pausing when you want to check what is happening to the population.
   a Look at what happens to the groups that we have discussed with time.
   b Keep an eye on the total population figure also.
3 Pause the animation when it gets to this year.
   a What age are the baby boomers now?
   b Why might that be a problem?
4 All figures after this are estimates. Play the animation to the end.
   a How does the pyramid change? What shape is it at the end?
   b According to the predictions, how many more people might there be in England and Wales in 2085 compared to this year?
   c What might happen to make these predictions wrong?

Dependency

By the end of this section you will:

- know and understand dependency rates and why these are important.

What is a dependency rate?

A dependency rate describes the proportion of people in a country that are dependent on other people in that country for support. For example, the population pyramids on page 115 show Belize (an LEDC) and Belgium (an MEDC). The pyramids show clearly that there are different proportions of the population in the different age groups.

Population geographers are most interested in three age groups:

- The youth-dependent group (0–14). This is the portion of the population which is not likely to be very economically active. That is, people in this age group are unlikely to be big earners, and it is likely, even in LEDCs, that many are still being supported by their families or by the older population.
- The independent or economically active population (15–64). This is the part of the population which is most likely to contain the wage earners and those who pay taxes to support schools, clinics and hospitals and building of roads and power stations.
- The aged-dependent group (65+). This is a group of the population which are likely to have stopped working and are perhaps starting to rely on pensions or other savings. This group is likely to need more and more social and health care as they age.

How can we calculate the dependency rate?

One way that the level of dependency can be calculated is using the dependency ratio, which is calculated by dividing the number of the dependent population by the active population and multiplying by 100. This is shown in the following formula:

$$\text{Dependency ratio} = \frac{\text{youth-dependent (0–14)} + \text{aged-dependent (65+) population}}{\text{working population (15–64)}} \times 100$$

Usually MEDCs have a dependency ratio of between 50 and 75. This means that for every 100 people of working age, they have to make enough money to support up to 75 dependents. LEDCs often have a dependency ratio of over 100. This is generally a large youth-dependency.

Aged-dependent population

- The aged-dependent population depends on the working population for pension contributions as well as health care and other social service benefits.
- In MEDCs with an increasingly elderly population, there is an increasing burden on the working population to pay for health care and pensions. As people have a longer life expectancy then pensions have to be paid for longer periods of time. Some governments such as the UK's are encouraging people to work beyond 65 so that they can support themselves for longer.

Figure 7 An example age–sex pyramid.

- While an elderly population is not a major problem for LEDCs at present, a growing proportion of older people in the future will have to be planned for. A previously high birth rate and improved health care will increase numbers in an LEDC's aged-dependent population over time.

Youth-dependent population

- The youth-dependent population is the youngest age group of the population. They are usually dependent on parents or guardians for food, water, clothing and shelter. They are also dependent on the independent population for education and health care, which is usually provided by the government from taxes paid by the working population.
- This population tends to be much larger in LEDCs than in MEDCs because of higher birth rates and larger proportions of young people in the population. However, in many LEDCs, as birth rates fall, the youth-dependency ratio is falling also. Countries with a high youth-dependency ratio have to provide a lot in terms of education and medical provision.

Independent or working population

- The working population, also known as active or independent, is those aged 15–64. It seems strange for us in Britain to consider 15-year-olds as in the working population, but for many parts of the world starting work at this young age is quite normal.
- The government collects taxes from the working population which are then used to pay for health care, education and social services such as child benefit or unemployment benefit. The working population may also be making pension contributions which will be used to support the population when they retire from work.
- The amount that is taken from the working population in taxes and pension contributions varies from country to country. The amount of money gathered in tax from each person in an MEDC country such as the UK is much higher than that which is provided in an LEDC such as Somalia. In LEDCs where there is a large working population there may not be enough jobs to go round. This leads to problems of unemployment, underemployment and an informal economy.

Research activities

1. Find the interactive population pyramid for Northern Ireland on the NINIS site (http://www.ninis2.nisra.gov.uk/public/InteractiveMapTheme.aspx?themeNumber=74 and open the link Northern Ireland Population Projections (1982–2064) – Population Pyramid). Working in a group, consider how the population pyramid of Northern Ireland is forecast to change over time.
2. Using precise figures and dates (hover over the bars on the interactive pyramid) describe how the population has changed, and is forecast to change in the future.
3. Give three bullet points to suggest possible consequences of those changes on each of the following:
   - services (schools, hospitals, care homes, etc.)
   - employment
   - education
   - pensions
   - society
   - the economy.

Activities

1. Find an animated population pyramid for China, a rapidly changing LEDC (in a search engine put the phrase 'Animated population pyramid China' and search – it should be the top link). It should show you the situation in 1950.
   - a What is the shape of the pyramid? Why is it that shape?
   - b Watch the animation – how does the shape change?
2. At the beginning there is a high youth dependent population.
   - a What happens to that over time? What else happens to the dependent population?
   - b In which year does China have the most economically active and the least dependent population? What happens after that year?

2 Causes and impacts of migration

An explanation of migration

By the end of this section you will:

- know and understand push and pull factors leading to migration
- know and understand physical and human barriers to migration.

Migration is the movement of people around the world. This has been happening since early humans first travelled from Africa, around 70,000 years ago, eventually to colonise the whole world.

Taking Ireland as an example, the present population of the island is 6.8 million. It is estimated that between 9 and 10 million people have emigrated from Ireland since 1700. These people were emigrants from Ireland, as this was where they left, and immigrants in the countries to which they moved. The destination countries included Great Britain, Canada, the USA, Australia, the countries in Central and South America and elsewhere. There are now said to be over 100 million people around the globe who have known Irish ancestors.

People can also migrate from areas that have been affected by changes in climate and extreme weather events. Additionally, those fleeing the civil war in Syria form just one example of the many migrations due to conflict.

What are push and pull factors?

A push factor is something that would encourage people to move away from an area, like unemployment or a natural disaster. Sometimes push factors only work for some people. A lack of opportunities for finding high-paying jobs might push one individual to move, but might be less important to another.

A pull factor is something about an environment that attracts people to move to a place, as in it would pull them away from where they currently live. People may be pulled to a place because of its climate, its economic prospects or because their children could go to good schools. A pull for one person, such as good salmon fishing or lots of nightlife, might not matter as much to another person, or might even put them off.

Push and pull factors can be personal to individuals. However, something like drought or famine might push many people out of an area, and safety and economic prosperity might pull a lot of people to places that offer those things.

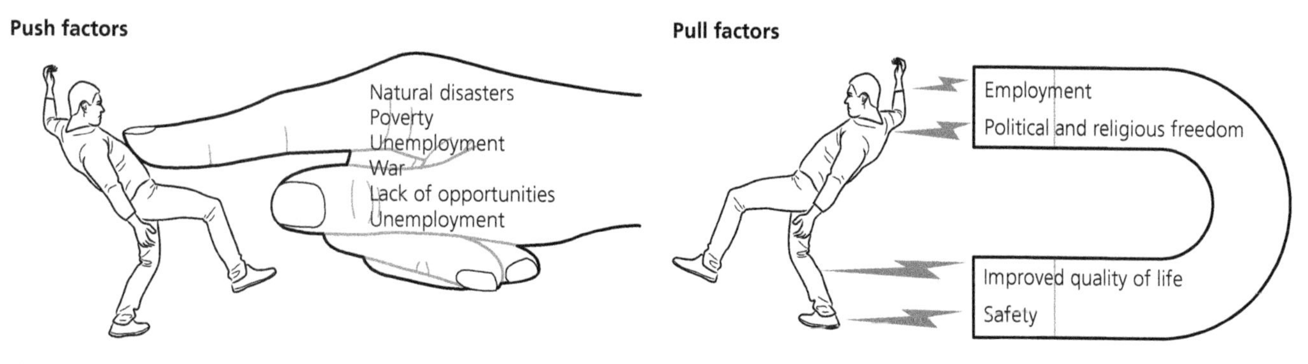

▲ **Figure 8** Push and pull factors.

What barriers are there to migration?

People do not always move because of push and pull factors. Sometimes this is because there are barriers to migration. These can be human or physical.

Human barriers are actions that people or nations take to prevent movements. Countries try to control movements by allowing only certain people in to their countries. Sometimes they require special permits, called visas, for anyone going to their country. These generally limit how long a person can stay in a country, and whether they can work while they are there.

Physical barriers include challenges to migration such as wide rivers, mountain ranges or deserts which impede movement. Some governments build physical barriers to try to prevent migration, as shown in Figure 9. These physical barriers can be overcome by those desperate enough to move to the other country, and modern transport has made some of them less important. However, migrants fleeing North Africa and aiming to reach Italy or Malta, or leaving Turkey to attempt to reach Greece, have to face the physical barrier of the Mediterranean Sea, and many have been drowned in recent years in their attempts to overcome this physical barrier.

➤ **Figure 9** The border between the USA (on the left) and Mexico (on the right).

Activities 

Work with a partner to examine the pushes and pulls on young people such as yourselves from Northern Ireland considering moving to the United States of America.

1. Do you agree on what might push you away from your home place and pull you towards the USA?
2. Are there some pushes and pulls that matter to you but don't matter to your partner?
3. Would these pressures to move be enough for you to leave home to travel to some distant place?
4. What would be the impact on Northern Ireland if a lot of young people decided to move away?
5. What barriers to movement might there be for you moving to the USA?

Types of migrants

By the end of this section you will be able to:

- distinguish between an asylum seeker and a refugee.

People who move far from one place to another are correctly called migrants. However, there are many different reasons why they move, and the type of migrant is determined by those reasons.

What are asylum seekers?

Someone who is fleeing from an area, perhaps because of war, drought or famine, is called an Internally Displaced Person (IDP) if they stay within their own country. If a person crosses a border to another country to seek protection or shelter, and to apply to become a refugee in that country, they are an asylum seeker – asylum is an old word meaning a hiding place or a place of protection and shelter. The Syrian communities that have been forced from their homes because of a brutal civil war who are living on the Syrian side of the Turkish border are not asylum seekers because they are still in their own country. If they were to cross into Turkey and did not want to return to their country of origin because they feared persecution due to their political views or their race, religion or social group, they could apply to be recognised by Turkey as refugees, and would then be defined as asylum seekers.

What are refugees?

When an asylum seeker goes to another country, they can formally request protection by claiming refugee status from that government.
At that stage they may become a refugee, which means they are protected from being sent back to their country of origin because of a proven fear of prosecution. Refugee status is recognised around the world as protection against being persecuted, although those fleeing from wars or droughts have less protection. Asylum seekers cannot work in the country to which they move, but if they are accepted as refugees, then they can look for work. Many refugees may eventually return to their original country, when conditions improve there. Some may wish to become citizens of the country that has protected them.

The United Nations High Commissioner for Refugees (UNHCR) was created in 1950 to help some of the millions of Europeans who had fled or lost their homes in the turmoil of the Second World War. Initially intended as a three-year project, it is still protecting and helping refugees around the world, and in 2015 had a budget of US$7 billion.

IDPs and asylum seekers, some of whom may eventually be recognised as refugees, are therefore people fleeing situations where they fear for their lives.

How are economic migrants classified?

Another category of migrant is an economic migrant. In this case the person is moving not to flee from persecution, but to improve their chances of getting employment and earning money. Economic migration has been the driving force behind many population movements in the past. Most of the movement of the Irish population to the USA, Canada, Australia, South Africa and New Zealand in the past was of economic migrants moving to other countries to seek a better life. Many Irish people still move to the USA to find work. These migrants are often young, and often single men, who find economic opportunities in their home countries limited and so they move to other countries to find work there. (Women sometimes face cultural barriers to work opportunities in their home countries, and it may be deemed more important that they stay at home, especially if there are children to look after.) Sometimes, if migrants find opportunities there, and they overcome any barriers such as language or racism, they can become established in the country to which they move, and may bring other members of their family to that country.

Economic migrants are often prepared to do jobs which others do not want to do. These jobs may be unpleasant, physically hard and poorly paid, perhaps working in food-processing factories or in gathering crops. These workers may also be forced to work in abusive conditions with poor wages, often below the legal minimum wage, long hours, and no sick leave or holidays. They may also be forced to live in crowded and poor quality accommodation. All of this is because some people see them as easy to exploit. Despite this, some people, in countries where there are economic migrants, may blame them for taking jobs from local people, which is seldom true. Others confuse asylum seekers or refugees with economic migrants. Migration was a key factor in the UK's 2016 EU referendum campaign, with many people complaining about misleading information being given out on both sides.

➤ **Figure 10** Migration was a key factor in the 2016 EU referendum campaign, causing many complaints about misleading information.

Activities 

Work with a partner to complete the following activities.

1. Find two or three definitions of an asylum seeker or a refugee online.
   a. Are they different? Are they in agreement? Which is the closest to being the correct definition?
   b. Agree on definitions in your own words of an asylum seeker and a refugee that would be understood by most people.
   c. Give three reasons why someone might become a refugee.
2. Do you know of any economic migrants in your family or friends?
   a. Why did they migrate?
   b. What were the main advantages for them, and what were the main disadvantages?
   c. Would you consider emigrating as an economic migrant at some stage in your life? What would encourage you to move, and what might hold you back?
3. People have different views about migrants, with some people welcoming them and others resenting their presence.
   a. Why do you think that people have the views that they do?
   b. What benefits do you feel that migrants bring to a country?
   c. Can you think of any problems that may result from immigration?
   d. If you became an asylum seeker, how would you feel in a different country?

CASE STUDY

Migrants to Greece, 2015–16

- After unrest in Syria had become a civil war in 2011, there was a large increase in the numbers of people who were internally displaced or became migrants, with many crossing the border into Turkey. The flow after the civil war began in Syria in 2011 is now one of the biggest movements of people into Europe ever documented. The flow was at first a trickle with around 8,000 migrants fleeing to Turkey in 2011. By 2014, there were an estimated 815,000 Syrian migrants in Turkey. They are still arriving. Of Syria's 22 million people, 6 million are internally displaced and 4.8 million have fled abroad, many to Turkey. Some of these migrants have tried to move on to European countries, many through Greece. While most of the migrants are Syrians, there are also people from Afghanistan and Iraq, many of them fleeing conflict in their countries. Poverty, human rights abuses and increasing violence have also persuaded many people to leave their homes in Eritrea, Somalia and other countries to make the dangerous trip to Europe, crossing the Mediterranean from north Africa. A quarter of those who made this hazardous journey were children. Many of the migrants move north as soon as they can, heading for countries such as Germany or the UK. However, more and more barriers are being put up to prevent this movement. Some of these are fences which physically prevent movement, such as along the southern border of Hungary, and other countries, such as Austria, have imposed caps on the numbers entering. This has meant that more and more people are stranded in Greece, which is a burden on that relatively poor member of the European Union.

What challenges do migrants in Greece face?

- In the first eight months of 2016, about 240,000 migrants, mostly from Syria but also Iraq and Afghanistan, made the perilous crossing to Greece from Turkey. There had only been about 40,000 during the whole of 2014. These migrants cannot move any further into Europe as many countries have closed their borders. Many are stuck at Greece's northern border, enduring increasingly unsanitary conditions. Others are in crowded camps that are not designed for the numbers which have to stay there. These are migrants who have escaped violence and war, but who continue to have concerns about inadequate food and health care. The Moria refugee camp, one of those on Lesbos, the main landing point for migrants on dinghies arriving from Turkey, has an increasing number of Syrians, Eritreans, Afghans, Pakistanis, Kurds and others. This camp is overcrowded with tales of families crammed in with 30 other people, including many babies, in tiny shelters fabricated from a shipping container.
- The EU struck a deal with Turkey whereby migrants arriving from there after 20 March 2016 were to be returned to Turkey if they had no documents, did not apply to be recognised as refugees or were refused refugee status. For every migrant returned in that way, the EU agreed to take one Syrian migrant from Turkey, to help the ease the pressure in that country. Greece must process each asylum application before the migrants can be transported back to Turkey or granted asylum. However, in the first six months after the deal was agreed, only 468 were returned, out of the more than 10,000 people who arrived. The bottlenecks have overwhelmed many of the camps, especially on the Greek islands, where migrants arriving after the March deal were supposed to be held prior to removal back to Turkey.

Figure 11 Migrant routes from north Africa to Europe.

What will happen next?

- The future for these migrants is unclear. Those who gain refugee status can move into the other countries that will accept them, such as Germany. The plan is that the others will be sent back to Turkey.
- The pace of dealing with this is slow. It is feared that these conditions will make crime more likely, or raise the number of migrants who might be tempted to join extremist movements. The inability of Greece to cope with the influx of migrants has meant that these people have not got a chance to settle, to send their children to school, to find jobs and to contribute to Europe's economy. As one migrant said, 'Our life is stuck. You have no job, no training, nothing.'
- The numbers of migrants and the camps set up to process them make this an unusual migrant process. Generally, migrants locate themselves in areas where they can gain employment, and they often live alongside similar people with the same language and customs, for reasons of feeling safe as a community and at home. They can make up sizable populations in their chosen location, although this can fuel resentment and racism. While in the cramped and unsanitary temporary camps, the migrants find it very difficult to form communities.

What challenges does Greece face?

- With high unemployment levels – standing at over 25 per cent overall and 50 per cent for youths – Greece has one of the weakest economies in the EU. On top of this, it is having to cope with this migrant crisis largely on its own. Living conditions and overcrowding in the camps have to be tackled but this is difficult as the stream of migrants continues, and there is no sign of an end to the civil war in Syria. Greece is effectively a holding pen for a growing number of people unable to travel further into Europe.
- Anti-immigrant sentiment is on the rise in many countries in Europe, and Greece is no exception. There have been racist attacks on migrants by supporters of Greece's neo-Nazi Golden Dawn party.
- There have been claims that the uncontrolled flow of migrants to Greece is putting pressure on Greek tourism and that is posing a direct threat to the national economy. Others have claimed that there has been no effect on tourism.
- While tourism may or may not be affected, there is an ongoing cost of keeping large numbers of migrants while the other countries of Europe decide how many Syrians they will accept as refugees.
- This unusual situation is unlike most other migrant population situations where a community puts down roots, its members building businesses, enrolling their children in local schools and finding work. Then the pressures on the host country are related to providing support for the migrant population, ensuring that they are integrated into the host community as much as possible. The situation in Greece is very different. Most of these people do not want to stay in Greece, where finding work is hard even for local people. Many of the migrants would return to Syria if the civil war ended. Others wish to settle with family members who are already established in Germany or the UK. Only when those people get there will they begin to be part of the expatriate Syrian community.
- The extraordinary movement of people from Syria has led to many acts of generosity by Greek people, ranging from small acts of kindness, like giving food and shelter to hungry and exhausted migrant families, to participating in more organised rescue operations for those at sea in unseaworthy boats.

Figure 12 An asylum seeker, holding his daughter, crying tears of joy as he arrives safely on the Greek island of Kos in August 2015.

Activities

1. Consider the position of migrants from Syria who are currently in Greece:
   a. List the impacts they are having on Greece.
   b. List the impacts that being in Greece has on them.
2. Figure 13 shows migration routes into Greece. Complete the following activities:
   a. Use an atlas to find the island close to the Turkish coast where most migrants have arrived.
   b. What are the other four islands in which they arrive?
   c. Where in the mainland of Greece are these migrants sent after they are processed?
3. Those migrants who do not want to wait for the slow process of allowing them to apply for asylum sometimes make the journey into the heart of Europe on foot. The Idomeni border crossing has more than 12,000 people in a camp designed for no more than a few thousand.
   a. Find a picture of the Idomeni Camp on the internet and put it in the middle of a page. Construct a spider diagram which shows the challenges faced by the migrants and the challenges faced by the country of Greece. Think about economic factors and social factors.
   b. What is missing in the camp, and what could the Greek authorities do to help?
   c. Imagine that you are in the Idomeni Camp. Write an email back to your family in Syria describing where you are, what it is like and what you plan to do.

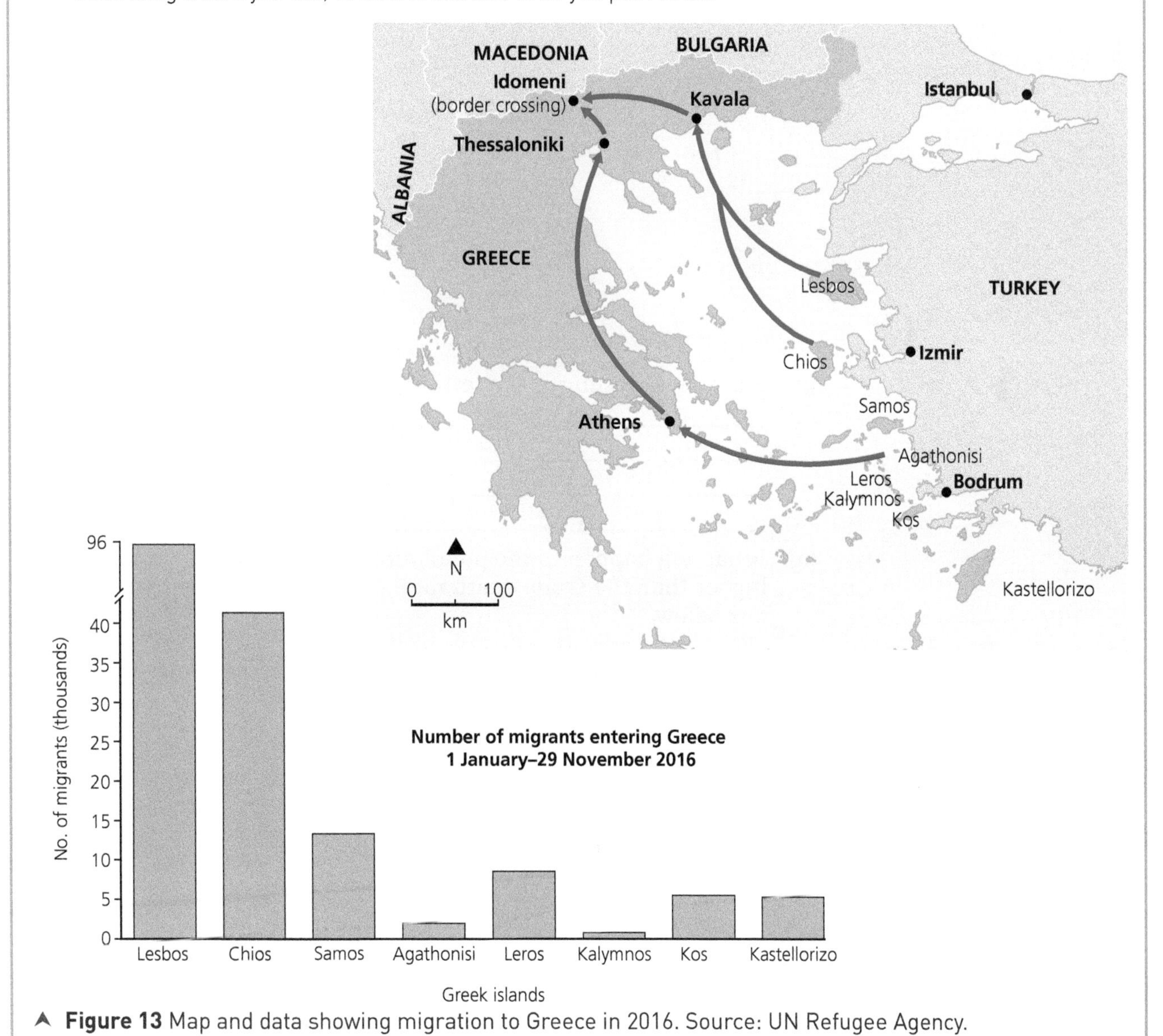

▲ **Figure 13** Map and data showing migration to Greece in 2016. Source: UN Refugee Agency.

Sample examination questions

1 Study Figure 1, which shows the crude birth rate for Africa, and answer the following questions.

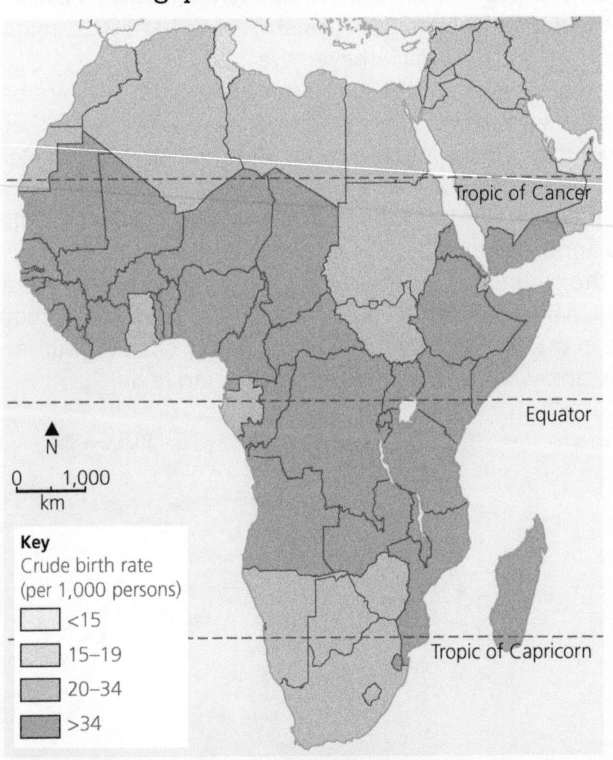

Figure 1 The crude birth rate in Africa.

(i) State the meaning of the term 'crude birth rate'. [2]

____

____

____

____

(ii) What will happen to the population size if the crude birth rate is higher than the crude death rate? Select from the options in the box below. [1]

Stay the same	Increase in size	Decrease in size

2 Study Table 1 and answer the questions below.

Country (continent)	Percentage of population		
	0–14 years	15–64 years	65 years +
Sweden (Europe)	17.3	62.6	20.1
Niger (Africa)	42.8	54.1	3.1
Ireland (Europe)	21.5	65.7	12.8
Nepal (Asia)	30.9	64.1	5.0

Table 1 Population structure in four countries in 2016.

(i) Select one MEDC country and one LEDC country and describe the differences in the population structure between them, as shown in Table 1. [4]

Name of LEDC chosen: Name of MEDC chosen:

(ii) Explain one **advantage** of a low youth dependency in some countries. [3]

3 Study Figure 2, which shows some factors which might encourage people to consider migrating, and answer the questions below.

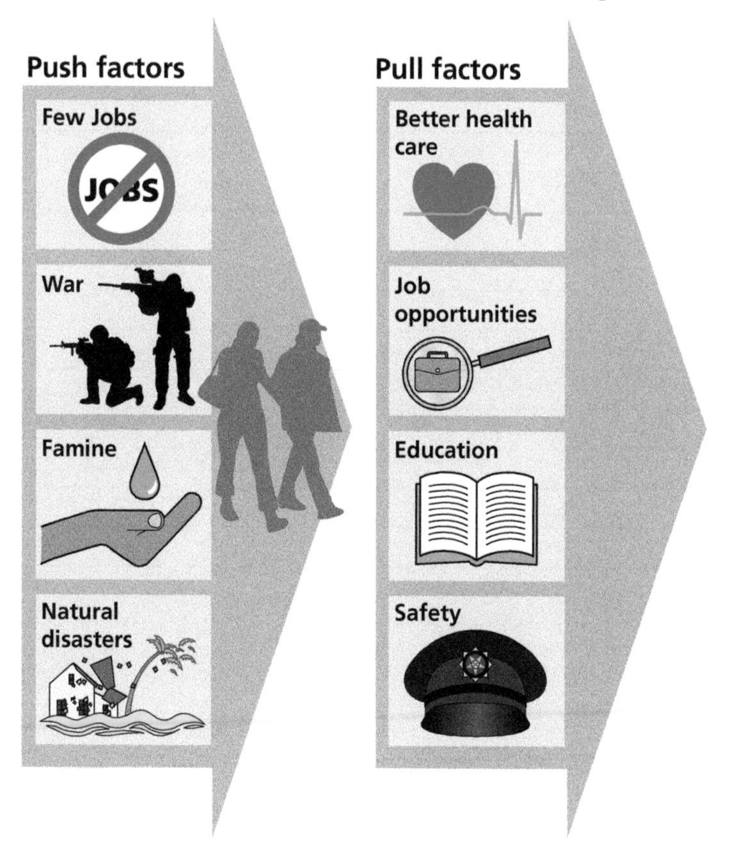

Figure 2 Migration push and pull factors.

(i) State the meaning of the term 'migration'. [2]

(ii) Using the diagram above to help you, explain the factors that encourage people to migrate. [5]

4 State the meaning of the term 'economic migrant'. [2]

5 With reference to your case study, discuss the challenges faced by refugees. [6]

UNIT 2

LIVING IN OUR WORLD

THEME B: Changing Urban Areas

Shanty town, Kolkata.

Why do many LEDC cities have shanty towns?

By the end of this section you will be able to:

- identify the characteristics and locations of different zones in a city.

The zones of a city

Larger settlements such as cities have areas with specialised land uses. This means that there are areas which have developed mainly as commercial, residential, industrial and so on. These are called functional land use zones.

Figure 1 Inner city terraces, Belfast's Holy Land.

What is a CBD?

At the centre of many cities is the central business district or CBD. This is where the main shops and offices are found, making it the centre for commerce and entertainment. All transport links lead to it so it is the most accessible part of the city. Because of this, it is the most desirable place for many different land uses. As the area that it covers is relatively small, and the demand for space in it is high, the rents that people would be willing to pay to locate there are the highest in the whole settlement. As a result, many properties are built upwards and so it is in CBDs that skyscrapers are located. For many people, the CBD is the settlement as it is the only place in the settlement that they visit. However, it is only a part of the settlement. With increased traffic the CBD can become very congested and this has led to streets being pedestrianised. Car parks are now found around the CBD and there are often 'park and ride' car parks located at the edge of the city to reduce traffic in the CBD.

Figure 2 The central business district (CBD) of Hong Kong.

What are inner cities?

Around the CBD in many cities is a residential area with a high population density. This is called the inner city and is often a place where there is urban decay and dereliction. It is a term generally used for places where the inhabitants are relatively poor, crime rates are high and unemployment is common. The inner city often has high levels of deprivation. Multiple-occupancy homes are common, where houses have been subdivided into flats. In some cities these areas may be in such a bad condition that they are called ghettos or slums. In other cases, parts of inner cities can become dominated with student accommodation (see Figure 1). This description of inner cities works for places like London, Toronto, Dublin or New York. However, in some other cities such as Paris, Rome or Sydney, the inner cities are the most prosperous parts of the settlement where housing is most expensive and where the richest people in the cities live. In those cities the poor live in the suburbs (see below).

Even in cities where the inner areas were places with the poorest quality housing and the most unemployment, changes can happen. Some parts of inner cities may become attractive to more wealthy people who start to live there again. As richer people move in, the character of the area starts to change, the location becomes more desirable, prices rise and the poor who were living there have to move elsewhere. Following the new people will be services such as upmarket supermarkets and delicatessens, restaurants and shops catering for this new population. Sometimes old buildings are taken over and converted to residential properties. These might already be residences or they may be old industrial buildings with some character. Sometimes new housing is built to cater for the wealthier incomers. This process is called gentrification and has taken place in many cities.

What are suburbs?

Suburbs are residential areas at the edge of settlements. They are characterised by detached single-family homes and low population densities. Most house people who travel to work in the CBD, although some suburbs are now areas of employment themselves as service and light manufacturing industries have grown up there. Residents of the suburbs tend to have a high proportion of car ownership, to allow them to move within the city. As most residents work in other land use zones, suburban streets tend to be quiet during the day.

Figure 3 Suburbs in east Belfast.

What are industrial zones?

Cities with an industrial base will contain industrial zones. The older industries may now be located in the inner cities as the city has grown around them. They are often associated with transport links such as railways or canals. They are likely to be in older buildings and will be surrounded by housing which originally would have accommodated their workers. This would be the case with many mill towns in Ireland and the UK. More modern industrial zones would be found on the outskirts. Here the buildings will be more modern and will rely on motorway transport to bring in raw materials and the workers and to distribute the finished product.

Figure 4 An old industrial landscape in Belfast of shipyards and terraced houses.

What is the rural–urban fringe?

Around the edge of many modern cities is an area which is not rural but not fully urban either. It is called the rural–urban fringe. Typically it is a place where recycling facilities are found and where park and ride facilities are provided. Airports and large hospitals are often located there too. However, despite these urban uses, the rural–urban fringe is largely open countryside although generally it is of poor quality with badly maintained hedgerows and woodland.

Urban sprawl is when a city or town spreads out into the surrounding countryside covering what was once farmland, woodland or wetlands with buildings such as roads, car parks, houses, shops and offices. This has a major impact on the environment as much of the wildlife is lost and water can no longer seep into the ground so easily, which can lead to flooding. Cities with extensive urban sprawl often have problems with transport, with much congestion and pollution caused by people travelling from the edge of a sprawling settlement into other zones for employment, leisure or shopping.

Figure 5 Rural–urban fringe near Belfast.

Activities

1 Taboo

Construct a game of Taboo for another group to play using the key words in this section:

- Urban sprawl
- Suburbs
- Gentrification
- Rural–urban fringe
- Inner cities
- Central Business District

On separate cards, write the key word and list up to five words which you might use if trying to explain the key word to someone else. These are the words that someone playing the game has to avoid using during the course of the game. You want to make it challenging but not impossible! Below is an example from Physical Geography:

Flood plain
- Flat
- River
- Deposition
- Fertile
- Risk of flooding

In this case the person explaining could not use flood (or flooded, or flooding, or floods or anything like that), river, flat, deposition, etc., while trying to explain what a flood plain is.

Your teacher will take the cards and pick one example of each key word. These will pass from group to group, as each has an opportunity to describe the key word within a time limit. If a group guesses what the word is, they get a point and so does the group doing the description.

Afterwards, think what skills were involved in this exercise – how might you use those skills in other contexts?

2 Street View

Use Google Earth™ to navigate to areas of a city which you know. Zoom in to parts of the city to identify the different zones that you would expect to find there, such as CBD, rural–urban fringe, suburbs, inner cities.

Make a screenshot to capture what the streets, the green spaces and the transport routes look like from an aerial view.

Then use Street View to get a sense of the place from the ground. Navigate around a little to get a representative view point and take a screenshot of that.

Match the Street View screenshot with the aerial view in each case. Do this for each of the land use zones that you would expect to find in a city.

Compare what you have done with the images found by others. Could you easily recognise the images that they have captured for CBD, rural–urban fringe, suburbs and so on?

By the end of this section you will be able to:

- interpret aerial photographs and OS maps to identify land use.

Using maps and aerial photographs

Looking at maps and aerial photographs can tell us a lot about what a particular part of the city does – its function. You can distinguish whether it is mostly for people to live in or to work in. Shopping areas or areas dominated by transport links can also be distinguished and there is a large shopping area, Connswater Shopping Centre, in the south-east of the map and photograph. Tall office blocks for workers providing services to others can also be seen in the south-west corner of the image and map.

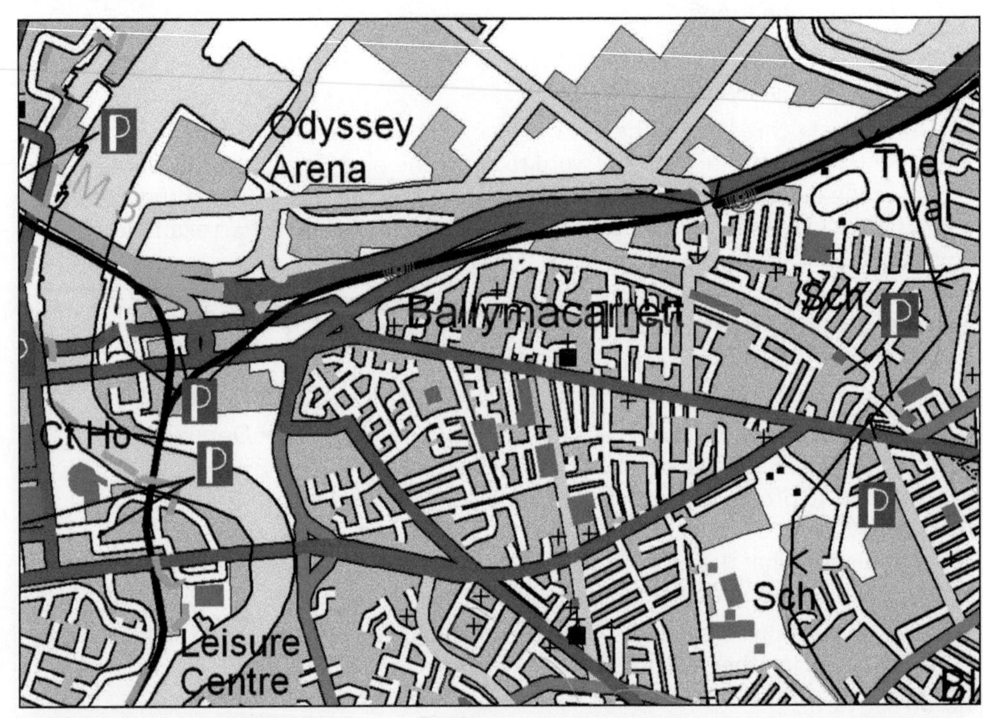

▲ **Figure 6** OS map extract of east Belfast.

▲ **Figure 7** Aerial photo of east Belfast.

How can we use maps and aerial photographs to look at settlements?

If you examine the map and aerial photograph of a part of Belfast (Figures 6 and 7) you should be able to see residential areas with lines of streets. Notice that, even though this is a landscape which is made by people, the streets still take account of some of the physical features such as the Connswater River, which they neatly avoid. The Connswater is the snaking river running south to north along the east of the photograph.

Look more closely at the photo to find any green space that could be used for recreation. The map also shows some recreational space, and some others can be inferred. The line of a former railway has been opened up for use by walkers and cyclists – the Comber Greenway. Other paths have been developed as well.

Heavy industries generally need a lot of space and houses nearby for their workers. Look for signs of heavy industries that are there now, or used to be there.

Residential, commercial, industrial or recreational functions are provided by a city and many can be identified from maps and aerial photographs.

Figures 8 and 9 show the rural–urban fringe at the edge of Derry/Londonderry. Look closely at the types of houses at the edge of the city and the size of their gardens. There is farmland to the east of the area, right at the edge of the city.

▲ **Figure 8** Aerial photo of Culmore Road, Derry/Londonderry.

Figure 9 Map of Culmore Road, Derry/Londonderry.

Activities

1. View the 1:50,000 Ordnance Survey map and the aerial photograph (Figures 6 and 7) alongside a digital map into which you can zoom (try www.openstreetmap.org) in order to identify places better.
   a. What do these areas look like on the maps, and on the aerial photograph? What are the main kinds of houses that you can see? What land use zone in the city would that suggest is being shown?
   b. There are also some factories. What do these look like? One of the biggest factories (Bombardier, found to the north of the map and aerial photograph) makes wings for aircraft, among other things. Why are factories located there? What do they need to keep them working? Why would that be a good location?
   c. Tall buildings are generally a feature of the CBD of cities. Are there any tall buildings in the aerial photograph? Look for large shadows. While just at the edge of the CBD of Belfast, this shows that details like this can be discovered from aerial photographs and maps as well.
2. Figures 8 and 9 show the edge of Derry/Londonderry, another city in Northern Ireland. The residences here are suburban and you can also see the edge of the countryside so this is the rural–urban fringe of the city.
   - In what ways do they differ from those in the image and map of Belfast? You can also see the rural–urban fringe. What is this zone and why is it there? How might it change if you were to take a new image 30 years from now? What might be some of the challenges for farming in that location, with the city right against the edge of your farm?

2 Issues facing inner city areas in MEDCs

Living in MEDC inner city areas

By the end of this section you will be able to:

- understand issues facing many MEDC inner city areas, such as:
  - housing
  - traffic
  - the cultural mix.

MEDC inner city areas are often vibrant and stimulating places to live. Being close to the CBD means that there is employment nearby, and many services such as cinemas, restaurants and clubs. There will be concert halls and museums, as well as parks and historic buildings nearby. The people that live in these areas can be from many different backgrounds and it can be exciting living alongside people with different cultures, different languages and ways of life. They may eat different food and celebrate different festivals from the main population of the city. If there is a university nearby, often these areas have a high student population which adds another community to the area. Figure 1 on page 130 shows a part of Belfast, the Holy Land, which has a lot of students, along with a mixture of other communities.

However, there are some challenges associated with living in inner city areas which are explored below.

What are the housing issues?

In many MEDCs the inner city is mostly composed of terraced housing, much of which was built to house workers in the nearby factories. The houses were often built over 100 years ago. When the areas were built, there were no cars on the streets and so there was no need to have garages or parking spaces. Bathrooms would not have been installed in these houses. The original factory workers and their families would have had a bath once a week in a tin bath in the middle of the floor. Many inner city housing areas would have had public baths. These would have had a swimming pool but also baths with hot running water, which people would not have had in their houses. There would typically have been a toilet, but it would have been at the end of the yard. The houses would have been heated with coal fires, but there was no central heating so a fireplace was needed in each room, including bedrooms. Look at the number of chimney pots in Figures 1 and 10.

Figure 10 Back-to-back houses in Birmingham, preserved as a tourist attraction by the National Trust.

Most of these houses were very well built, often of red brick. The houses that were not have been torn down, as slum housing was cleared in the 1960s. The ones that were torn down were often back-to-back houses with a shared courtyard with a communal toilet block for the whole housing area. Often they had no back door. In some places these have been preserved so that people nowadays can see what they were like (see Figure 10).

Most of the terraces that remain have been improved, with bathrooms being added and the toilet being moved indoors. Sometimes the two ground floor rooms, the front parlour and the kitchen, have had their dividing wall removed to make a larger ground floor room. However, the houses are still cramped, there is little space around them and there may be very little green space nearby. For families with children, there is limited play space and the streets may be busy with vehicles. The corner shops which would have been scattered about these communities have now largely closed due to pressure from supermarkets, and this makes it difficult for elderly people or those without access to a private car to go shopping, use the Post Office and so on.

These terraced houses once housed very strong communities with neighbours looking out for each other and children playing together in the street and in each other's houses. While there still are some strong communities, most have less strong ties as new residents move in and the original residents become elderly or move out of the area. There is also higher unemployment in such areas now.

The houses, while relatively solidly built, are over a century old. They do not have cavity walls – a gap between the inner and the outer walls of the building – so they may need renovation to prevent damp. The electrical systems and fixtures such as windows and doors may be in need of replacement. The roofs are generally made of heavy slate, and they too may need regular attention to keep them from leaking. Older houses are often draughty and very expensive to heat, as they will generally be very poorly insulated, and this makes them inefficient.

Gentrification of residential areas

A process that has been taking place in many inner cities in MEDCs has been gentrification. This is the process where people with higher incomes buy properties and renovate them to live in themselves. This has advantages and disadvantages, some of which are below.

Some local firms, such as the fishmongers near Brixton Market in London shown in Figure 12, may find it difficult to continue in business. This could cause a closure of many family-run firms and the city as a whole may be poorer and less interesting as a result.

Those families forced out by higher rents or by lack of access to the services that they need often have to relocate to housing estates in the suburbs. This can cause friends, families and communities, who previously lived close together, to lose contact with one another.

Advantages of gentrification	Disadvantages of gentrification
The area can be improved in appearance with more wealthy residents improving the properties.	The sense of community can be undermined for the original residents.
Smart shops, cafes and bars often open with the new influx of wealth.	The original residents would prefer a small corner shop and do not value, or cannot afford, services like coffee bars.
Old buildings are preserved and maintained.	The price of renting or buying properties increases as the area becomes gentrified, pricing local people out of the area.
The streetscape may be improved with the planting of trees and provision of new facilities.	The differences between the wealthy newcomers and the relatively poor original residents may lead to resentment.
New businesses, often multinational chains, move in.	Small businesses and low-income families move out.

Figure 11 A gentrified property.

Figure 12 Gentrification in London.

What are the traffic issues?

Inner city areas were built before most people had cars and the main mode of transport was walking, or taking public trams or buses. Now the areas around the CBD in most cities is full of cars. As the inner city is the zone next to the CBD, with all its shops and offices, workplaces and places of entertainment, people who bring their cars in to use the CBD will often try to park in the terraced streets close to the centre. Many of the local residents will also have cars, but these terraced houses were not built with garages, nor do they generally have the space to build any. The result is increased congestion in the inner city. This has negative consequences for air quality and journey times.

Figure 13 Pollution in Mexico City.

The impact of congestion on air quality

Air quality in cities like London is so poor it is illegal. Most of the pollution is caused by diesel vehicles which may meet emission levels in tests in the lab, but emit much more when out on the roads. The greatest problem is nitrogen dioxide (NO_2), a pollutant which is invisible and has no smell. Nitrogen dioxide affects the lungs of those who breathe it. The lungs of children are stunted, affecting those people for life. Additionally, exposure to NO_2 increases the chances of getting asthma or lung cancer. It is not just pedestrians; even those driving the cars are affected. The government estimates that this pollutant leads to the premature deaths of 23,500 people across the UK each year, and 70,000 across Europe.

London established a Low Emission Zone in 2008. This charges the most heavily polluting lorries and vans entering this area. However, there is some evidence that this has made little difference to the levels of pollution. There is a plan to establish an Ultra Low Emission Zone (ULEZ) in central London by 2020, although London's mayor

'London certainly has significant pollution, enough to have effects on health. It is a hidden killer.'
Prof Sir Malcolm Green, founder of the British Lung Foundation.

'Those populations most exposed to air pollution are also more likely to have other stressors, such as poverty, poor housing, low educational attainment, obesity, long-term illnesses and higher levels of smoking.'
Royal College of Physicians.

'Children are ultimately defenceless. They can't vote but they are lumped with the health effects for life.'
Simon Birkett, director of the Clean Air in London campaign group.

'Children, older people, and people with chronic health problems are among the most vulnerable to air pollution.'
Royal College of Physicians.

'We need to ban diesels as we banned coal 60 years ago.'
Simon Birkett, Clean Air in London.

Figure 14 Opinions on air pollution.

Sadiq Khan has plans to make the ULEZ larger and bring it in in 2019.. Again, this will work by charging drivers of the more polluting vehicles. The original ULEZ plan had been criticised as it only covered 300,000 people in the very centre of London, and not the 3 million living in highly polluted inner London boroughs around that centre. A report by the Policy Exchange found that a Londoner's life expectancy is shortened by about 16 months by air pollution. It also highlights that poorer neighbourhoods – including many of those inner city areas – are the worst affected.

In 2016, the mayors of four cities – Paris, Madrid, Athens and Mexico City – undertook to ban all diesel cars and trucks by 2020, following Tokyo, which did so in 2003. Diesel vehicles produce less carbon dioxide, a greenhouse gas, than petrol cars, but produce four times the amount of nitrogen dioxide. They also produce particulates: tiny particles which have a negative effect on health. London may soon follow the lead of these cities. However, as there are 11 million diesel cars on the UK's roads, any outright ban would be difficult.

The impact of congestion on journey time

London has about 20 per cent of all of the UK's traffic congestion. Congestion can be measured as the extra travel time in minutes per kilometre compared to what you might expect if there was no congestion and the traffic moved smoothly. In central London, the congestion during the morning peak is an average of 1.4 minutes per kilometre. To put this into perspective, if travelling at 30 miles per hour (48 kph), the usual speed limit for an urban area, you would expect to take 75 seconds to travel one kilometre. Instead, on average, it would take 2 minutes and 40 seconds for each kilometre, travelling at an average of around 14 mph (22.5 kph). Congestion is rising in London and traffic is moving more and more slowly.

The reasons for this increase are unclear. Increases in delivery vans and minicabs have been suggested, for example. Whatever the reason, there are now more vehicles using London's roads than ever before, and some road space has been given over to other priorities such as cycling. Construction work is also a constant part of London's continued growth, and this can increase congestion further.

London's congestion is estimated to cost at least £4 billion per year, as longer and more unreliable journey times affect business productivity and harm the economy. Congested roads also make the air pollution worse, increasing the detrimental impact on Londoners' health and quality of life. In addition, the road safety of vulnerable users such as cyclists, pedestrians and motorcyclists is also threatened by congestion.

How can a person living in an inner city reduce the effect of air pollution?

- Monitor where the worst pollution is, perhaps by using social networking. Avoid these places.
- If you have respiratory or heart problems or are elderly, you should avoid strenuous exercise on days when pollution levels are high.
- As road traffic causes most air pollution, avoid busy roads and junctions, especially if there are high buildings and little wind. Side roads are best.
- Air pollution can be a third lower on the inside of the pavement, compared to closer to the traffic.
- Car drivers and passengers are affected as well. Walking and cycling expose you to pollution but this is still healthier than using a vehicle.
- If you need to be in a vehicle, keep the windows closed and recycle the air in the car rather than venting air from the outside into the vehicle.
- Don't exercise in heavily polluted places as fast breathing causes more pollution to reach your lungs. Running, for example, is best done away from main roads and when traffic is quiet.
- Diet is also believed to help protect against pollution, and studies in Mexico City have found that children who eat more fresh fruit and vegetables were better protected against some pollutants.

Public transport

While inner city residents may often have to put up with congestion and air pollution from traffic, they often do not own a vehicle themselves. Instead they have to rely on public transport. Public transport is much better for the environment than using cars (see Figure 15). However, relying on public transport can be expensive.

It might also be less efficient. London only started running tube trains all through the night in October 2016, and only on the Central and the Victoria lines at first. If you worked in central London during the night, for example as an office cleaner, there might have been little choice in public transport available to get you to work or to get you home again.

Using public transport is expensive in the long term. One report, although mostly about commuters from poorer parts of the edge of London, says that London residents who earn more than £600 per month have to work 20 minutes every day to pay for their commuting. Those earning between £200 and £599 have to work for 54 minutes to pay their commuting costs, and those earning less than £200, which made up 26 per cent of all workers, have to work for 1 hour and 56 minutes before they have paid off the cost of their journey to work each day. In London, 21 per cent are paid below the London Living Wage (£9.75 per hour in 2017, compared to the National Living Wage of £8.45 per hour in 2017, and the National Minimum Wage of £7.05 for those aged 21–24). There are an estimated 940,000 Londoners living in poverty in the inner city, although this has fallen as poor people are forced to move to the edges of the city.

▲ **Figure 15** Buses are more efficient and better for the environment than cars. There are as many people travelling in this bus as in the queues of cars beside it.

Parking

The cost of parking in inner cities is also an issue, due to a combination of limited space and efforts to curb pollution. Often this cost is so high that it forces poor people to use the relatively expensive option of public transport.

Efforts to combat this include:

- the building of multi-storey car parks that allow a lot of cars to park in a small 'footprint'
- resident parking permits allowing only local residents to park
- freeways, where no parking is permitted to allow the flow of traffic.

Some MEDC cities such as San Francisco, Los Angeles and Stockholm are moving towards smart-parking solutions, which combine low cost sensors that can locate an empty parking spot with smart parking meters. Drivers can be informed of an available parking space through an app and be navigated to it. This avoids drivers having to circle areas to find a parking place, reducing pollution and congestion. There is a cost to this solution, but the Los Angeles city authorities were able to pay off their investment in just three months.

A more radical solution is to dissuade drivers from taking their cars into city centres at all. City authorities can prioritise less-polluting transport solutions by building bus lanes or cycle paths, or by pedestrianising whole streets. They might also develop park and ride schemes, which encourage drivers to leave their cars in the suburbs. Some cities have very ambitious targets: Oslo wants to ban all private transport in its city centre by 2019.

What are the cultural issues?

The inner city areas of MEDC cities often comprises a rich mix of nationalities, languages, foods and a wide array of traditional dress. This is a diverse population, often made up of people who have only recently made their way to the country and who congregate in areas where accommodation is relatively cheap, and work is as close as possible. These areas are often where others from the same background have settled, and this can be comforting to newcomers to a city and country which perhaps seems quite unfamiliar to them, and in which they may experience discrimination. London is one of the most ethnically diverse cities on the planet, with many ethnic groups, races and languages. All of these communities can add a lot to a settlement by enriching and enlivening these areas. One example would be the Notting Hill carnival in London. Led by the British West Indian community, this street festival attracts around a million people each August and is a celebration of London's multicultural diversity.

Ethnic tensions

Figure 16 Muslim women wearing niqabs.

While there are many benefits from the cultural mix, there can also be some problems. Even smaller, less ethnically diverse cities such as Belfast are not immune from ethnic tensions between the host community and recent arrivals. More than 100 Roma people left their homes in Belfast in 2009, out of a total population of just 200, when there were a series of attacks on them. There was also some trouble at an international Northern Ireland vs Poland football match in the same year, and Polish flags were burned in 2011 and 2012, to the concern of the 30,000 people of Polish origin living in Northern Ireland and many local people.

Religious tensions

There can sometimes be religious tensions in multicultural communities. Despite London electing the first Muslim mayor in a major European capital city in 2016, there are continued tensions. Hate crimes against Muslims have been on the rise. Women are more often the subject of attacks, perhaps because their clothes often mark them out as Muslim, and they may be seen as easier targets for abusers. It is believed that the numbers are much higher, but that those attacked often fail to report the incidents in case it would make matters worse.

Year	Number of hate crimes against Muslims in London
2013–14	478
2014–15	816

Source: Metropolitan Police

Language barriers

It is believed that there are over 300 languages spoken in London, the capital of the United Kingdom. Over 100,000 people speak Bengali and Silheti, which are Bangladeshi languages; Panjabi, originating from the Punjab region in India and Pakistan; Gujerati, originally from India; and Urdu from Pakistan. However, there are speakers of Russian, Hebrew, Japanese, German, Somali, Cantonese, Albanian, Swahili and many other languages as well. Big cities such as London attract a lot of people to work, and this brings a vibrant mix of nationalities. However, there can be some disadvantages. Access to health care can be a problem if the individual has little English, and the signs and patient information are only in English. Those with limited English may find it difficult to get access to any state services, or employment, and may find it harder to fit into the wider community. Generally these individuals are highly motivated and hard-working, and they are keen to learn English, with all the benefits that brings (though older people in particular may struggle to learn a new language).

Memo

You and your team have to describe the problems of an inner city to a set of urban planners who want to understand the area.

You have to:

- Produce a digital presentation with six slides which present the problems of the area
- Provide images on each of your slides and figures where possible. You should also have a cover slide.
- Make sure the presentation lasts no longer than 5 minutes – these planners have busy schedules. Be prepared to answer any questions that they might have.

The company is on show here – do a good job!

Activities 

1 Work in a team to take on the brief shown in the memo on the right. Present your brief to the other groups.
- Do not:
  - read the text from the slides.
- Do:
  - work as a team in preparing the presentation and presenting it
  - keep as much eye contact with your audience as possible
  - speak confidently
  - describe each slide and what it shows clearly
  - keep to your time limit.

2 Read this newspaper report.
- Work with a partner to decide what you would spend the money on, to try to improve this inner city area. What would improve life most for the people of this area? What would be sustainable? What might attract young people away from a life of drugs and gangs? What might give them a better future?

3 Put a photograph of an inner city area in the middle of a large piece of paper (or do it digitally using a computer).
- In one colour draw lines from the photograph out to boxes around the image and provide details of all the good things about living in an inner city.
- In a different colour, draw lines and provide details about the challenges of living in an inner city.
- Give figures where possible.

The north-east inner city of Dublin has many problems, many of which are related to poverty and deprivation. The census figures from 2011 showed that there are pockets of male unemployment of over 70%. One third of the area is classified as 'areas of disadvantage' and 10% is 'extremely disadvantaged'. The roots of these economic problems stem from the 1960s and 1970s when the two main employment opportunities, docking and clothes manufacturing, died out and there were no replacements.

There are many social problems. Over 50% of the housing in north-east inner city Dublin is rented from the council. In some pockets, lone-parent households make up 80% of all households. There is a high percentage of young people who leave school before they are supposed to, often with no qualifications. Drugs are a problem, particularly for young people, and children as young as 12 or 13 have been known to get into debt because of them.

While there are pockets of poverty, there are many areas of the north-east inner city which have a lot going for them, and involvement in sport has helped many young people. The vast majority of the people are law-abiding and hard-working.

In 2016, north-east inner city Dublin was promised €1.6 million in funding to tackle some of the problems there, particularly related to gang-related violence and drugs. Additionally, a garda (police) station which had closed was going to be reopened and a 10-year vision for the area was announced.

By the end of this section you will be able to:

- evaluate one MEDC urban planning scheme that aims to regenerate and improve housing, employment, transport and the environment.

The Titanic Quarter, Belfast: MEDC urbanisation

What is urban planning?

Urban planning is when people try to shape settlements, or towns, so that they work more effectively. Universities offer courses in town planning and it is often a career chosen by people who have studied human geography. Increasingly, urban planning schemes are incorporating **sustainability** in their decisions. If cities are going to meet not only the needs of the present residents, but also the needs of future residents, settlements must be developed sustainably.

How has urban planning regenerated and improved Belfast?

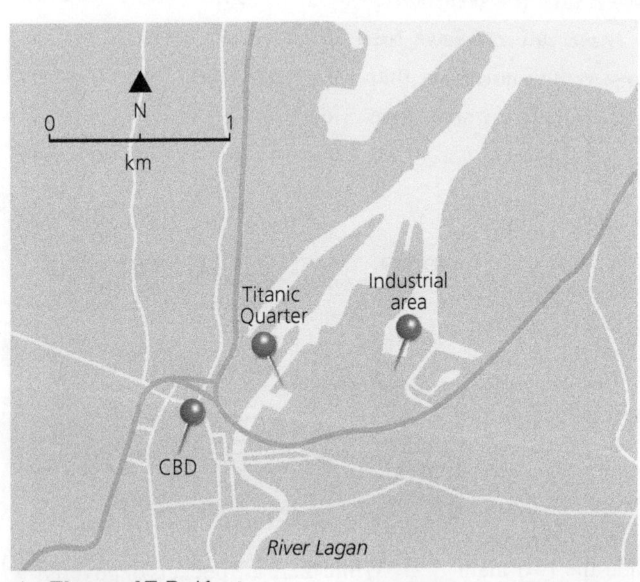

Figure 17 Belfast.

Figure 18 Regeneration of the Titanic Quarter (TQ).

Belfast is a city that has relied on industries – such as spinning and weaving of linen, tobacco processing, shipbuilding, heavy engineering and rope manufacture – for its prosperity. For many generations, the River Lagan was vital to the people of Belfast. It was the focus of trade and of industry, and the port at the mouth of the Lagan was where goods were exported and imported. This trade made Belfast a powerhouse of the Industrial Revolution, the period of rapid growth in technology and populations, around 1750–1900. This port at the mouth of the Lagan was also the place where ships were built and launched. In the 1800s Belfast boasted the world's largest shipyard (Harland and Wolff), and the largest rope works and linen mills in the world.

Since the peak of Belfast's industrial power, the city has largely turned its back on the river. Riverside areas have become derelict and polluted, and many of the people who lived in these areas have moved away. The shipyard declined as large passenger ships and tankers began to be built in other parts of the world, such as in Asia.

However, Belfast shipyard is a brownfield site. Building on brownfield sites reduces pressure on greenfield sites, as fewer new developments on open land are needed. This is one of the reasons why the company Titanic Quarter Ltd chose to make this 75-hectare area of inner city Belfast into an exciting waterfront development, called Titanic Quarter (TQ). By 2016, £358 million had been invested in this urban planning scheme, and 18,000 people were living, working or studying in TQ.

Employment: leisure and tourism

TQ draws many visitors to its growing list of attractions. These include the T13 Urban Sports Park, restaurants and cafes, a marina and the adjoining Odyssey complex, with its 10,000 seater SSE Arena and the W5 discovery centre, all of which employ local people.

At the heart of the development is *Titanic* Belfast, celebrating the shipyard where the RMS *Titanic* was built. First opening in 2012, Titanic Belfast attracted over 2 million visitors from 124 countries in just four years, bringing in around £105 million for the Northern Ireland economy. This development employs many people with, as their website says, 'a wide range of work disciplines and skills areas'.

Other employment

TQ already offers a large number of workspaces (139 thousand m^2), and there are plans for many more in a further 3 million square feet of office space. A number of large companies such as banks and accountancy firms have already moved to offices in TQ. Any modern urban development needs advanced and secure network connections, and these are provided in TQ. Connections are in every workspace and there is a direct Optical Fibre link to North America: the fastest from anywhere in Europe.

Some industrial buildings such as the Paint Hall, once used to paint parts of ships, have had a new lease of life. This building, now called the Titanic Studios, is one of Europe's largest film studios attracting big budget film-makers with sets for *City of Ember*, *Your Highness* and *Game of Thrones*, and providing further employment.

Housing

The first residential developments in TQ are now occupied. The ARC is a complex of 474 apartments with ground floor retail outlets. Plans are in place to build a further 5,000 apartments on the site. The developers believe that the convenient access to the centre of Belfast, and the range of activities in TQ, will be attractive to potential residents, as well as the 'extensive local amenities and high quality public realm spaces'.

Transport

TQ is very close to the central business district (CBD) of Belfast and may, in the future, become an extension of it. This makes it easily accessible for work, living and for leisure, particularly with public transport. There are dedicated bus services, but the area will also be built to allow rapid transport systems to be added later, if they are developed in Belfast. Pedestrian walkways and cycle routes will also be part of the development.

Figure 19 Titanic Belfast, a visitor centre which explores the story of the *Titanic* through galleries and community activities.

Environment

Most jobs and entertainment venues in cities are in the centre while the workforce lives around the edge of the settlement, or outside it. This results in long commuting times, traffic congestion and increased pollution. Living, working and 'playing' in one area, such as in TQ, reduces some of these problems.

Construction materials

Some of the area is heavily contaminated as a result of industry in the past and this will be restored. Locally available, recyclable, reusable and environmentally responsible materials will be used in the construction, where possible.

Energy and climate change

All of the buildings in TQ will be constructed so as to reduce carbon dioxide emissions as much as possible. Careful design and the use of energy conservation measures such as energy efficient light fittings, rainwater harvesting and solar heating will ensure this.

Biodiversity

TQ is located in an important coastal site in Belfast Lough. The company aims to support and improve biodiversity by creating landscaped areas and habitats.

Is the Titanic Quarter scheme achieving its aims?

The story so far

The TQ project is expected to take at least 30 years to be established, so it is still 'early days' for judging its success. So far TQ has been successful in attracting housing development and employment. The relocation of the Public Records Office to TQ, costing £30 million, and the building of a £44 million campus of Belfast Metropolitan College has given a boost to the area. The opening of the first hotel on the site, a Premier Inn, in 2010 also helped. The Science Park, employing 1,500 people, has boosted employment opportunities in the area. Titanic Studios also provides much needed employment. Titanic Belfast, which cost £97 million to build, had by 2016 attracted over 3 million visitors. It was also named the leading visitor attraction in Europe in the 2016 World Travel Awards, beating off the Acropolis in Athens, Buckingham Palace in London, the Eiffel Tower in Paris and the Colosseum in Rome. TQ claims that there are currently 15–18,000 people working, studying and living there. The fact that all of these developments have taken place during a period of economic instability, following the global financial crisis of 2008, makes them more impressive.

Plans for the future

There are proposals for more office accommodation, and a Financial Services Campus is planned, which will bring more employment. Ultimately it is intended that TQ will provide homes and jobs for 50,000 people.

Ongoing transformation for TQ

The TQ scheme was designed to regenerate this former industrial area, and to improve housing, employment, transport and the environment. Overall, there has been an investment of £385 million to do this. Undoubtedly, the area has been transformed. However, this investment would have to continue if TQ were to continue to grow and prosper. TQ lists the benefits of the area to investors, emphasising the workforce in Northern Ireland as one of Europe's youngest and fastest growing with 60 per cent of the population of working age and with labour turnover of less than 8 per cent. The benefits of high-speed IT **infrastructure** are also promoted prominently. However, another of the benefits, 'direct access to EU markets', is now less certain following the UK's decision to leave the European Union. There is tremendous potential for a successful TQ at the heart of Belfast and Northern Ireland, but efforts to promote it will have to continue.

Research activities 

1 Develop a pitch for a potential investor in TQ. You will need to provide facts that will attract the investor and persuade them to invest in the area.

2 Use the historical imagery in software such as Google Earth™ to explore the TQ site to see how it has changed over the years. You may want to view street level data to see the buildings using Street View on Google Earth™ or Mapillary (www.mapillary.com), to get the most recent images. Describe what you see, using the following questions to guide you:
- Is the area fully developed?
- Are there challenges?
- Is it being transformed?
- What might it look like in 20 years' time?

▼ **Figure 20** Photographs of the TQ before and after the development.

Pump house for filling and emptying the dry dock, now a tourist attraction.

The drawing offices where the Titanic was designed, now a 4* hotel.

Dry dock next to the pump house.

View of Sampson (l) and Goliath (r), the iconic shipyard cranes.

The Marina and some of the residential units.

The *SS Nomadic* which ferried passengers out to the Titanic, now a tourist attraction.

By the end of this section you will be able to:

- describe and explain the location, rapid growth and characteristics of shanty towns in LEDC cities.

LEDC Urbanisation

What are the characteristics of shanty towns in LEDC cities?

Almost a billion people, or 32 per cent of the world's urban population, live in slums, most of them in LEDC cities. These people live in overcrowded areas of housing, known as shanty towns, with very few services such as piped water, street lighting and so on. Usually the shanties are situated on land which is of limited value, perhaps at the edge of a railway track or canal or even land on which it is dangerous to build.

The waste from the shanty towns not only is not removed, it builds up in the area and affects the health of the inhabitants, particularly that of the children who live there. Residents of shanty towns have more health problems and less access to education, social services and employment than other urban dwellers. Most have very low incomes.

The number of slum dwellers is projected to increase to about 2 billion over the next 30 years if action is not taken.

Shanty towns have grown up for a number of reasons, depending on the history of the city in which they are found. Most are a starting point for people who have just moved to the city, perhaps from the countryside, or for people who are temporarily in financial trouble. They provide a place where people can live cheaply for a time. Most residents of shanty towns aim not to stay there but to make enough money to afford a better place to live. While a number succeed, a high proportion do not. Life is difficult and uncertain for shanty town dwellers, particularly those who are not in stable employment but who work in the informal sector, a precarious existence. This sector is not like formal employment but may include trading at street corners or hawking goods to lines of traffic. Shoe shining or other occupations are also part of this informal sector of employment. The residents of many shanty towns have no legal right to live on the land on which their shacks are built and this makes their life insecure as they could return to see their houses bulldozed at any time. Often the residents live in poverty and despair.

CASE STUDY

Kolkata, India

- Kolkata, formerly known as Calcutta, is a rapidly growing Indian city with a population which is not known exactly but is probably about 14.1 million (2011 data). It grew up on flat swampland on either side of the River Hooghly, part of the Ganges Delta. There are three areas of poor-quality housing in Kolkata:
  - In the centre of the city are poor-quality houses dating back up to 150 years which grew up as the city urbanised.
  - Around industries and at road intersections and alongside roads and railways are shanty towns which grew up 60 years ago.
  - More modern shanty towns are found on previously unoccupied lands, especially to the east of the city, and along roads and railways.
- In 2001, there were 5,500 shanty towns in Kolkata, of which only 2,011 were registered. The census of 2011 found that 29.6 per cent of Kolkata's population lived in slums. Slums are defined by the Indian Government as places 'unfit for human habitation', and yet over a quarter of the people in this enormous city lives in such places.
- Registered slums are called bustees in this part of India. They are recognised by the city authorities, and the people who live there have a right to live on the land they occupy. While bustees are the best quality of Kolkata's slum properties even they are built of poor-quality materials and are tightly packed together.
- The residents of unregistered slums are much worse off. They have no rights to occupy the land on which their shacks are built, often along the sides of roads and canals or on other vacant land. The quality of these houses is very poor.

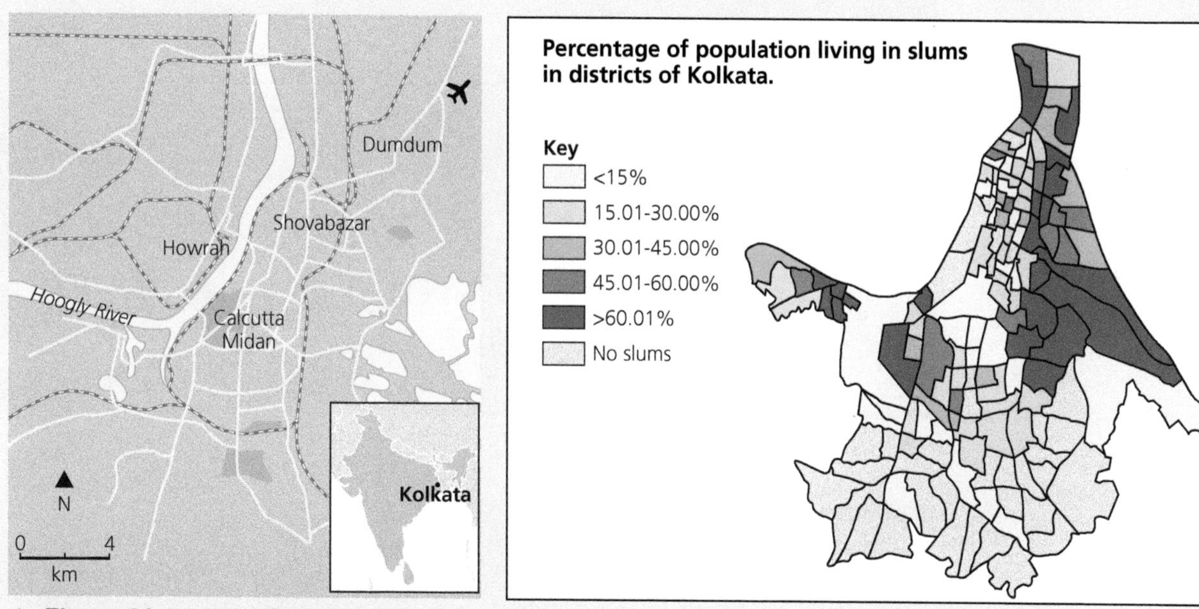

Figure 21 Areas in Kolkata and the percentage of slum dwellers in each part of the city.

The people living in the shanty towns

- Over 40 per cent of the inhabitants of shanty towns in Kolkata have been there for two generations or longer. Most of them have migrated into the city from the surrounding countryside, partly pulled by the prospects that a large city would seem to offer and partly pushed by low wages and mechanisation in the farms. Most inhabitants of the shanty towns work in the informal sector and have an average household size of five or six people. Their average monthly earnings are between 500 and 1,700 rupees (between £7 and £24), so about 75 per cent of the Kolkata shanty town population are below the official poverty line.
- The number of poor people who live in these shanty towns is set to increase considerably in the near future because of natural increases in the urban population and also because of in-migration.
- The standard of living of the shanty town residents was a cause for concern even during British colonial

rule. The slums were seen as a nuisance and an eyesore which threatened the security and the health and hygiene of the city's rich population. Successive governments tried to deal with them by removing them but each time this was done, they just sprang up again. Since 1974 the emphasis has changed to trying to improve the living conditions of the slum dwellers. This scheme, called Environment Improvement in Urban Sector (EIUS), has been partially successful in the shanty towns where it was put into operation. However, the continued rise of population in the city has led to new slums being built.

- Despite some success, Kolkata has a long way to go to solve the problems of the bustees and the unregistered slums. Clear long-term strategies are needed to reduce poverty, involve local people in developments, deal with the problem of unauthorised new slums growing up and provide further improvements to bustees. The problem is of such a scale that it would challenge an urban planner anywhere in the world.

Slum dwellers of Kolkata	Number/proportion of people
Illiteracy rate	71.5%
No sewage disposal	85.2%
No regular or steady income	22.2%
Actively seeking work	34.0%
Proportion of deaths due to diseases of digestive system	26.1%
Living in the slums for over 30 years	41%
Average family size	5.5 people

Figure 22 Statistics of the Kolkata shanty towns.

Figure 23 A shanty town in Kolkata, India.

Activities 

1. Using the details on Kolkata's shanty towns, work with a partner to identify a number of key facts about shanty towns in Kolkata. Source a range of digital images showing the shanty settlements in Kolkata and use a digital presentation to place each fact on top of each image, matching them as much as possible. Aim for at least six slides, but no more than ten.
2. Find a map showing shanty towns in a LEDC city.
   a. Describe where they are found. This will be different depending on the place you have chosen – what will be common is that the shanty towns will be built on land not wanted by others. Perhaps it is too marshy, too steep, too likely to be washed away, too dangerous because it is close to cars or trains or some other reasons.
   b. Find out the characteristics of your chosen shanty town. Try to get some description of it and some images. What is it like? Who lives there? How do they make a living? Where have they come from? Why do they stay in the shanty town? What is their quality of life?

Sample examination questions

1 Match the urban land use with the correct description by drawing a line between them. [4]

Urban land use
Inner City
CBD
Rural-urban fringe
Suburbs

Description
Terraced houses
Detached houses/housing estates
Mixture of land uses
Tall buildings

2 Inner cities in MEDCs can suffer from many problems. For each of the following, briefly describe the problem.

(i) Poor quality housing [2]

(ii) Traffic congestion [2]

(iii) Public transport [2]

3 (i) State the meaning of the term 'shanty town'. [2]

(ii) For a shanty town you have studied, describe and explain its location. [3]

Name of the city in which it is located: ____________________

4 (i) Describe a MEDC urban planning scheme that you have studied. [4]

Name of the urban planning scheme: ____________________

(ii) Evaluate how effective it has been in terms of housing, employment and environment. [6]

UNIT 2

LIVING IN OUR WORLD

THEME C: Contrasts in World Development

▲ A homeless man outside a car showroom in Kolkata, India.

Why has globalisation increased and not decreased inequalities in countries?

1 The development gap

By the end of this section you will be able to:

- identify the differences in development between MEDCs and LEDCs
- evaluate indicators of development, specifically the Human Development Index (HDI).

Differences in level of development

Geographers are interested in differences in levels of development in different countries around the world – how much more developed is the UK than Burundi, a small African country, and why are there such differences? They are also interested in variations in levels of development within countries – how much more developed is the south-east of England compared to the Highlands of Scotland, and what are the reasons which would explain the differences?

How is development measured?

There are many ways in which development can be measured, and each indicator of development has its own advantages and disadvantages. In the past, development was measured in economic terms only, usually in gross national income (GNI) per person. That is the total that a country earns every year, converted into US$ for easy comparison, divided by the population of that country. In 2015, the UK's GNI per person was about US$43,340. Monaco had the highest GNI per person in the world (US$186,950) while Somalia had the lowest GNI, at just US$150. This is an enormous development gap, measured in economic terms.

Based on wealth, countries can be divided into the rich, industrialised 'north' (MEDCs), including Canada, the UK, France, Ireland, Germany and Russia; and the poorer, economically less-developed 'south' (LEDCs), such as Sierra Leone, Nepal, Malawi, Niger, Bangladesh and Peru.

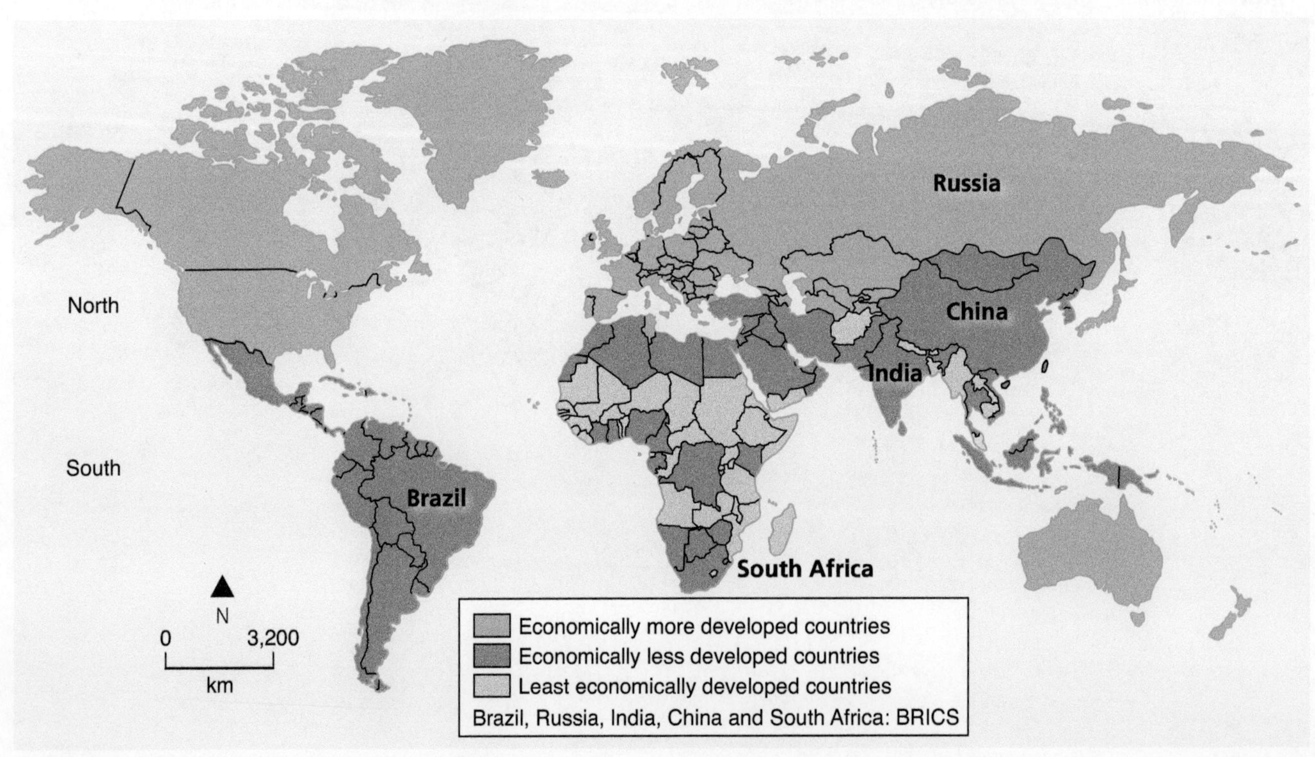

Figure 1 The global north–south divide.

Figure 1 shows the north–south divide which illustrates the economic difference between the rich 'north' and the poor 'south'. Note, however, that this is a generalised map. It places countries which are industrialising rapidly, such as India and China, into the same category as countries with little industry. Enormously wealthy countries such as the oil-rich Arab states are also classified as this. Some countries in the rich north on this map, such as Albania, or Moldova, are quite poor but are in the same category as the USA and Germany. The other thing that the map does not show is that rich countries can have poor regions within them, and poor countries can have rich areas.

What about social and economic indicators?

Measuring development by wealth alone does not explain all the differences in the world. Some countries with enormous wealth might not be considered 'developed' in other ways. To measure development in this way alone reduces development to just one variable – economics. Increasingly, it is seen as important to consider other facts too when measuring development, especially indicators of the social welfare of the population. Northern Ireland, for example, would be far below the south-east of England in most economic measures, with lower earnings, higher unemployment, fewer job opportunities, and less spending money, for example. However, not everyone in Northern Ireland wants to move to the south-east of England, and that may be because local people feel that there are some better social factors which keep them in Northern Ireland. Perhaps that would explain why, in 2014, Northern Ireland was voted to be the happiest place to live in the UK. Additionally, County Donegal, a relatively poor county in the Republic of Ireland, topped the *National Geographic*'s list as the world's 'coolest destination' for 2017. Measures of the quality of life have to be more than just about wealth.

Indicators of development can be divided into social indicators, which relate to people's well-being – they measure human welfare, and economic indicators which measure wealth. Figure 2 shows some socio-economic indicators of development for a range of countries. It suggests that there are clear links between wealth and a range of social factors. MEDCs have a longer life expectancy, a slower natural population increase, higher literacy rates and fewer people per doctor than countries that are economically less developed.

Country	GNI per person (US$) 2015	Life expectancy at birth (years) 2014	Annual population growth (%) 2015	Urban population (%) 2011 (*2015)	Adult literacy (%) 2015 estimates (*2007 estimates)	Doctors per 1,000 of population 2011 (*1995)	Computers per 1,000 population 2004 (*2005)	Internet users (%) 2016	Human development rank 2015
Ireland	46,680	81	0.5	62.2	99.0*	1.7*	494	81.0	0.916
UK	43,340	81	0.8	79.6	99.0*	2.8	600	92.6	0.907
Canada	47,550	82	0.9	80.7	99.0*	2.2*	700	88.5	0.913
Italy	32,790	83	0.0	68.4	99.2	4.1	367*	65.6	0.873
Brazil	9,850	74	0.9	90.6	92.6	1.9	105	66.4	0.755
China	7,820	76	0.5	56.1*	96.4	1.5	41	52.2	0.727
Lesotho	1,330	50	1.2	27.6	79.4	0.2*	n/a	20.6	0.497
Bangladesh	1,190	72	1.2	28.4	61.5	0.4	12	13.2	0.570
Nepal	730	70	1.2	31.0	64.7	0.04*	4	17.2	0.548
Sierra Leone	630	51	2.2	39.2	48.1	n/a	n/a	2.4	0.413

▲ **Figure 2** Socio-economic indicators of development (source: World Bank and UNESCO).

Key indicators of development

- **GNI per person:** this is the gross national income – all of the money earned by a country each year – divided by the population of the country. In Figure 2 this has been adjusted according to how expensive each country is and how far earnings might go (purchasing power parity).
- **Life expectancy:** the number of years that a person is expected to live for, on average, at birth.
- **Annual population growth:** the rate at which the population is growing, calculated by subtracting the death rate from the birth rate. A minus figure would indicate a falling population.
- **Urban population:** the proportion of the population that lives in towns or cities.
- **Adult literacy:** the percentage of people aged 15 or over who can read and write.
- **Human Development Rank:** the position of a country on the human development table produced by the United Nations (see the most recent at http://hdr.undp.org/en/data).

Problems with indicators

One major problem with social indicators is that the information is often obtained from a census or household survey. Most countries carry these out, but only every ten years or so as they are very expensive. It is very difficult to get information about every person in a country, even when the country can afford to spend a lot of money on a census. For less wealthy countries, or for countries where some of the population live in remote areas or moves around a lot, it is very difficult to get accurate figures.

In addition, some social indicators are related closely to the wealth of a country. A country with a wealthy population is likely to have more people with televisions or access to the internet.

The use of separate indicators can also disguise variations in development. For example, when comparing GNI per person, countries with a high proportion of farmers who grow food for the use of their family, and exchange some of it for other things that they need (subsistence farmers), will seem to be worse off than they are, despite their population being well fed.

For these reasons, some composite measures have been developed by the United Nations Development Programme (UNDP). Composite measures combine a lot of different indicators. The main one is the Human Development Index (HDI).

The Human Development Index

This index is expressed as a figure between 0 and 1 (see Figure 3). Countries can be ranked according to their HDI score. The closer the score is to 1, the higher the Human Development Index, and the more 'developed' the country is. The index combines measures of health, wealth and education, as follows:

- Life expectancy at birth, measured in years.
- Adult literacy rate and the combined primary, secondary and tertiary education enrolment.
- GDP per person, measured in US$ (GDP is gross domestic product, similar to GNI).

The benefits of a composite index are that there is less emphasis on just one feature, such as wealth. Combining a measure of wealth with social measures means that a more accurate measure of development is produced. The measures in the composite index have to be quite easily measured, even in relatively poor countries. However, even though the HDI is accepted as a useful composite index, it could be argued that other measures could usefully be added. Some geographers feel that the HDI still puts too much emphasis on wealth and suggest other factors such as freedom of speech should be included.

For many countries around the world, the HDI is improving, indicating increased living standards and a better quality of life. A number of countries in Sub-Saharan Africa (partly because of HIV/AIDs) and the new countries of Central Asia, for instance Kazakstan, are not seeing their HDI rise, due to worsening education provision and high mortality rates. These countries are the ones which stand out when the data are mapped as in Figure 3.

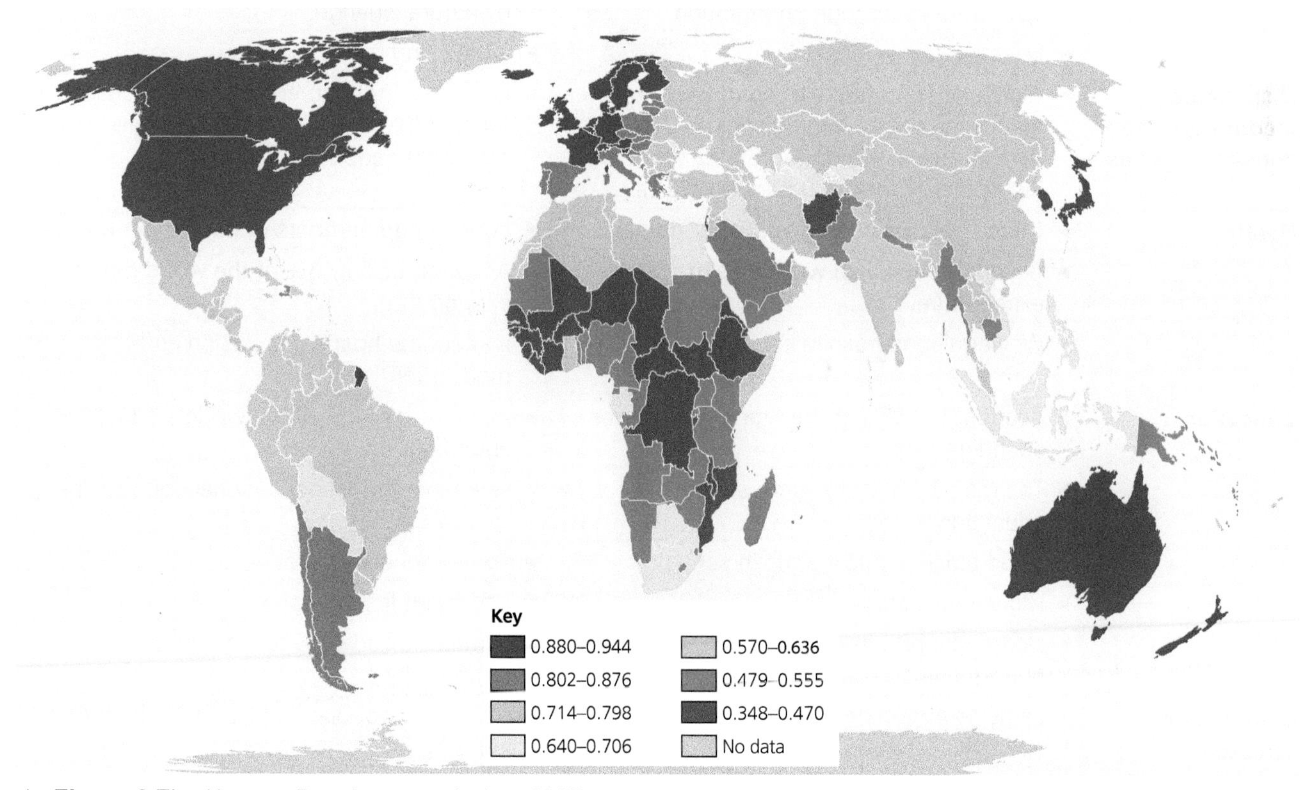

▲ **Figure 3** The Human Development Index, 2015.

Evaluating the indicators

In conclusion, it can be seen that there are vast differences between countries in terms of their development. Whether it is measured in terms of income, or in social measures or in a composite of different measures, the gap is enormous and is unsustainable environmentally, economically or politically.

Gustave Speth of the United Nations Development Programme describes sustainable human development as: 'development that not only generates economic growth but distributes its benefits equitably [fairly]; that regenerates the environment rather than destroying it; that empowers people rather than marginalising them. It gives priority to the poor, enlarging their choices and opportunities, and provides for their participation in decisions affecting them. It is development that is pro-poor, pro-nature, pro-jobs, pro-democracy, pro-women and pro-children.'

▼ **Figure 4** Differences in development between MEDCs and LEDCs.

	MEDCs	LEDCs
Gross national product (GNP)	▶ The majority have a GNP per person of more than US$5,000 per year. MEDCs have 83% of the world's income.	▶ The majority generate less than US$2,000 per person each year. Collectively, LEDCs have only 17% of the world's income.
	▶ High percentage of population is above poverty line; that is, they have an income of more than US$14.4 per day.	▶ High percentage of population is in extreme poverty; that is, they have an income of less than US$1 per day.
Life expectancy, population and population growth	▶ Over 75 years.	▶ Under 60 years.
	▶ 25% of world population.	▶ 75% of world population.
	▶ Relatively slow growth, for example Sweden's population grew by 0.79% in 2014, mostly through immigration.	▶ Fast growth, for example Burundi's population grew by 3.3% in 2014, mostly by natural change.
	▶ Effective family planning.	▶ Little or no family planning.
Disposable income spent on consumer goods	▶ High – large numbers of consumer durables purchased due to large disposable income, for example in the United States there are 797 vehicles per 1,000 people (2014).	▶ Low – few consumer durables purchased, for example in Bangladesh there are 3 cars per 1,000 people (2010).
Health	▶ Good – relatively few people per doctor.	▶ Poor – large numbers of people per doctor.
	▶ Account for 94% of world health expenditure.	▶ Account for only 6% of the world health expenditure.
	▶ Well-equipped hospitals.	▶ Inadequate hospital provision and medication.
Education	▶ Account for 89% of the world's education spending.	▶ Account for 11% of the world's education spending.
	▶ The majority have full-time secondary education.	▶ Few have formal educational opportunities.
	▶ Good teacher–pupil ratios in schools.	▶ Poor teacher-pupil ratios.
	▶ High adult literacy rates.	▶ Low adult literacy rates.
	▶ No gender bias in educational opportunities.	▶ Females disadvantaged in educational opportunities.
Employment structure	▶ Large % of population involved in secondary and tertiary industry.	▶ High % of population involved in primary industry.
	▶ 75% of world's manufacturing industry.	▶ 25% of world's manufacturing industry – much of it Multinational Corporation owned.

	MEDCs	LEDCs
Levels of technology/ mechanisation	▸ Highly mechanised.	▸ Mostly manual labour and animal power.
	▸ Large investment in research and development.	▸ Little local investment in research and development.
	▸ 92% of world's industry.	▸ 8% of world's industry.
Diet/access to clean water	▸ Balanced diet but increasing obesity.	▸ Much malnutrition.
	▸ High animal protein diet.	▸ Low protein diet.
	▸ 70% of world's food grains consumed.	▸ 30% of world's food grains consumed.
	▸ Majority of population have access to clean water.	▸ Many people do not have access to clean water.
Energy	▸ High levels of consumption, use 75% of world energy.	▸ Lower consumption rates, currently 25% of world energy.
Communications	▸ Good communications infrastructure: roads, railways and airports.	▸ Communications infrastructure focused on urban areas. Limited elsewhere.
Exports	▸ 82% of world's export earnings.	▸ 18% of world's export earnings.
	▸ Mostly manufactured goods.	▸ Based on primary products and unprocessed raw materials.

Activities

1 Work in a group to list nine measures that could be used to suggest differences in levels of development between countries. Write these on adhesive coloured square paper.
   a Arrange them into social or economic measures.
      - How many of them are social measures?
      - How many are economic?
      - Are there some which are difficult to say? Why is that?
   b Arrange the pieces of paper into a diamond nine with the measures that are best at measuring development at the top, the worst at the bottom and three in the middle.
      - Why is the one you have chosen as the best measure good at measuring development?
      - Why is the one you have chosen as the worst measure not so good at measuring development?
   c Take your three top measures and put them on a wall alongside the top three from the other groups. Is there any agreement? Are there some surprises? Talk to the other groups about why they made their choices, and explain why you made yours.
   d Working with a partner, write or paste a printed copy of the Gustave Speth quotation on page 158 into the middle of a large sheet of paper. Surround the quotation with facts, and photographs which illustrate what he says. Aim to get at least nine boxes around the quote. Compare your final product with those produced by other pairs.

2 Working in groups or as a whole class, each person picks a country and writes this on a card. You each need to find out the following about your country (you can use the CIA World Factbook website to do this – use a search engine to find it):
   - GNP (US$) per person
   - Death rate (per thousand)
   - Life expectancy (years)
   - Human development rank
   - Population
   - Birth rate (per thousand)
   - Adult literacy rate

   a When you have completed your card, stand in a line alongside the others in the class with, at one end, the highest figure and at the other end the lowest figure for the first measure of development (GNP). Who has the biggest GNP; who has the smallest? Go down the line hearing each country's name.
   b Now move to line up according to birth rate. Does the order of the line change much?
   c Work through all the measures, discussing the possible reasons for why you get the movement that you do.
   d Write a paragraph saying what this activity tells you about development that you didn't know before.

By the end of this section you will:

- understand some of the factors that hinder development in LEDCs.

Factors that hinder development in LEDCs

We may know what development is, and something about how to measure it, but what causes contrasts in development levels in the first place? There are a number of reasons which explain why development in LEDCs is limited.

What are the historical factors?

Colonialism was the system in which many Western European countries such as the UK, France, the Netherlands, Germany, Spain, Portugal and Belgium took over the running of vast areas of land elsewhere around the world from the sixteenth century onwards, and took resources and wealth from them. This included Britain forcefully enslaving inhabitants of the colonised countries and transporting them across the Atlantic Ocean to be made to work in plantations in the Caribbean and in mainland North America. All of the areas exploited by these European powers became colonies of the richer countries, and those in South and Central America, Africa and Asia provided raw materials which allowed the colonising country to develop their industries and become richer. At the same time the colonies had their industries destroyed and they became poorer. It suited the colonial powers to keep the employment in their countries and sell the goods to the colonies. India, for example, grew cotton, but cotton weaving was banned. Instead, cotton was sent to mills in England and the cotton cloth sold back to India and others. Colonialism as a system ended between the 1940s and 1970s. For example, Malaysia became independent from the UK in 1957, Algeria achieved independence from France in 1962 and Mozambique became independent from Portugal in 1975.

The lands that had been colonised were often divided into countries by the colonial powers with little thought about the people who lived there, and often international borders were decided in Europe by people who had never visited the area. Many of these boundaries were drawn in such a way that people who had a history of distrust or violence against each other were brought into the same country; others divided groups who had always lived together and who now found themselves in different countries. After the countries gained independence they often found that they had different groups competing for power, and deep social divisions.

Figure 5 Gold diggers.

The development of the colonised countries might have been helped a little by colonialism, and sometimes a good network of roads and railways and an efficient education sector may have been established by the colonial powers, for example. However, colonialism had mainly a negative effect on the colonised countries, leaving them dependent on the markets in the MEDCs even after they became independent. They also lag behind after a century in which little development was allowed to take place. The roads and railways and the education system were often constructed in order to allow the wealth of the colonised country to be exploited largely for the benefit of the colonising country. People who believe this often point to the fact that MEDCs continue to get much of their wealth as a result of LEDCs being kept poor.

What are the environmental factors?

Natural hazards such as droughts, floods, hurricanes, earthquakes and volcanoes often hinder economic development. Disasters such as floods can make it difficult for countries to expand economically. On the other hand, floods bring rich sediments, and volcanic areas are also very fertile. Of course, natural hazards are not restricted to LEDCs. There are volcanoes, floods, droughts and earthquakes in many MEDCs also, and hurricanes sweep the southern coast of the USA each year. However, LEDCs are less able to afford to build defences for these hazards and they often lack the finances to reduce the effects for their populations once the disasters have happened.

There are also some diseases which thrive in tropical climates, where many of the LEDCs are located. Malaria was once common in Europe, but it was eradicated with the draining of marshes where the mosquitoes which spread the disease breed. The disease remains a major problem in Africa, particularly because the illness restricts how effectively someone can work. HIV/AIDS too is a problem for many African countries, although rich countries have contained the disease in their populations with drug treatments. Many African communities do not have access to such health care.

The environment in many LEDCs is also under pressure from economic activities. Mining, forestry and, more recently, tourism can all put the environment at risk. This is particularly the case when attempts are made to access larger and larger profits. For example, deforestation earns a lot of money for a country but, when the trees have been removed, the land is vulnerable to desertification. Rainforests cleared for cattle ranching to produce beef are vulnerable to the leaching of minerals because of the high rainfall, and the land left behind may have little agricultural value as a result. The clearance of mangroves to farm prawns along many coasts in South East Asia makes those coastal areas, which previously the mangroves protected, vulnerable to storms.

What is meant by dependence on primary activities?

Most employment in MEDCs is in tertiary activities – work in services such as teaching, nursing, banking, refuse collection, banking, catering and so on (see Figure 6 on page 162). A very small percentage of people in MEDCs are employed in primary activities – work in extractive industries such as mining, farming, forestry and so on. LEDCs are very different, although that is changing. At present they generally have poorly developed tertiary industries. They often have very high proportions of their populations involved in primary industry and can depend heavily on this for much of their earnings. For example, Zambia depends heavily on copper for 85 per cent of its exports – almost all of its income as a country.

This is a problem for these countries as the prices paid for the primary products produced – commodities such as cotton or bananas or copper ore – vary enormously from year to year. The result is that the countries find it very difficult to plan ahead. If you want to build a motorway or a new hospital, you have to be sure of your earnings. Many LEDCs are not sure from year to year. Figure 7 shows how copper prices varied over a nine-year period. This makes it very difficult for Zambia's Government to plan ahead and means it is very vulnerable to a collapse of the international price of copper.

The ownership of the primary activities is also challenging. You are used to seeing vegetables from Peru and Kenya, seafood from Thailand and Vietnam and evidence of other foodstuffs from other LEDCs in your supermarkets. You don't see where the iron ore used to make your washing machine comes from, but it is also likely to be from LEDCs. However, the mining companies which exploit the minerals, and the agricultural conglomerations who work the land for commercial farming in LEDCs, are almost always owned by MEDC corporations. The country in which the mine or the rich farmland is located does make money, but most of the profits go to rich companies in rich countries. Zambia exported a total of US$10,000,000,000 (US$10 billion) in 2011, mostly copper, but Zambia itself collected only US$240,000,000 in tax mining revenue. While that is a lot of money for a poor country, it is just 2.4 per cent of the value of the country's minerals that are sold abroad.

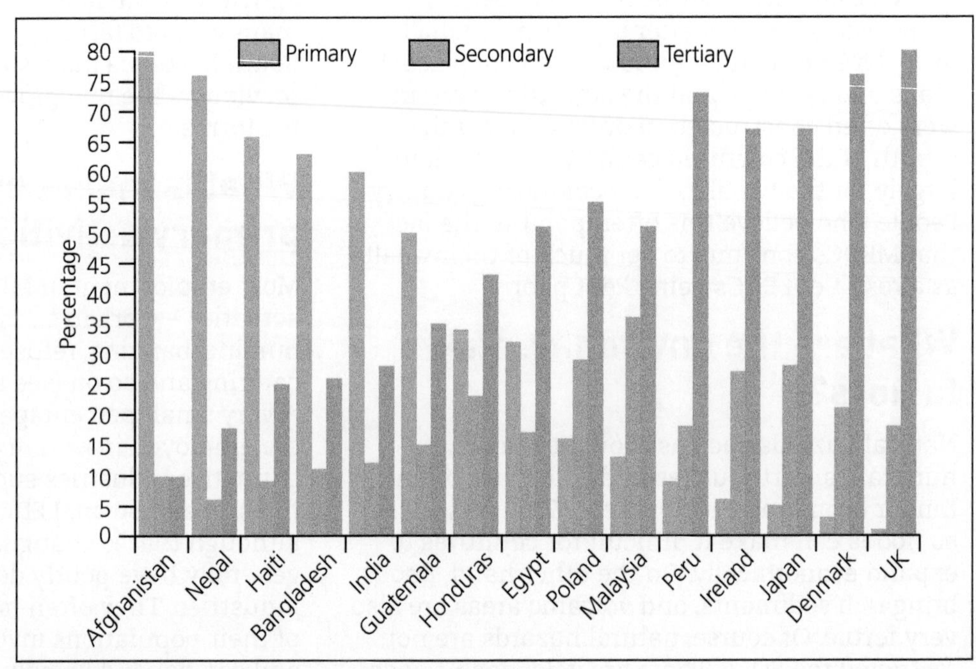

Figure 6 Types of economic activity (2011).

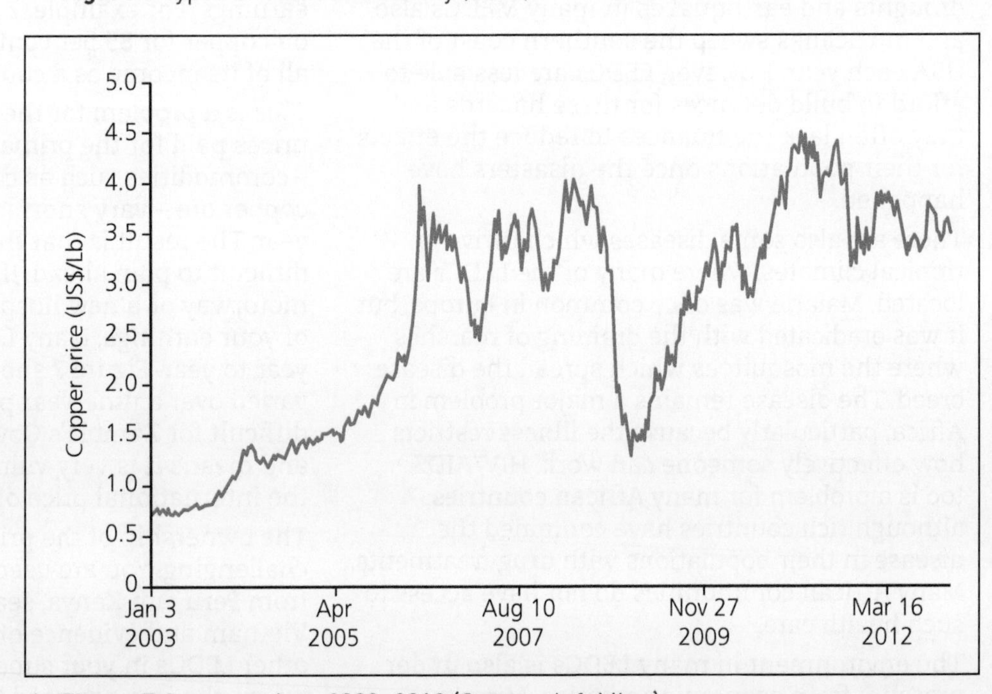

Figure 7 Copper prices 2003–2012 (Source: InfoMine).

Why is debt a factor?

Following the end of colonialism, banks in MEDCs lent enormous amounts of money to many LEDCs. Much of the money was to be spent on massive infrastructural developments such as dams, airports and roads. This was thought to be a good way to stimulate development.

Inevitably, the scale of the lending was such that many countries had difficulties in repaying the loans.

One example should indicate what happened. Ecuador is a tiny country in South America, on the equator, hence its name. In the 1970s the Government of Ecuador borrowed US$3 billion from international lenders. The government at that time was a dictatorship and spent much of the money on the military, to ensure that it retained power. Ecuador has been a democracy since 1979 but the debt still needs to be paid. As interest is charged on the loan, soon Ecuador owed more than it borrowed in the first place. By 2015, the debt was more than US$34 billion, having risen by US$2.6 billion since the previous year. There is a tremendous human cost to this growing debt as every dollar spent paying it off means a dollar less for fighting poverty and improving the quality of life of the people of Ecuador. In 2007, Ecuador paid US$1.75 billion towards its debt, which is more than it could afford to pay for social services, health care, the environment and housing and urban development put together. In December 2008, the Government of Ecuador threatened that it would not pay back the money that it owed, with the President calling the debt 'immoral and illegitimate'. However, the debt continues to grow, even with the vast payments made each year.

Here are some startling facts: the total debt of LEDCs continues to rise, despite ever-increasing payments, while some aid is falling. For example:

- The poorest 60 countries have spent US$550 billion in paying off both loans and the interest on the loans over the last 30 years. The original debt totalled US$540 billion; however, there is still US$523 billion to be paid. These figures are so high it is hard to understand the scale of the problem. What we can say, though, is that the countries have paid off more than they borrowed but, because of interest on the loan, owe almost as much as when they took out the loan in the first place.
- The G8 countries (France, Germany, Italy, USA, Japan, UK, Canada and Russia – although Russia has been suspended) agreed to cancel US$100 billion of the debt following pressure from Make Poverty History and the Jubilee Debt Coalition (Drop the Debt).
- The Heavily Indebted Poor Countries (HIPC) initiative involves providing support to 38 LEDCs with high levels of poverty but high debts. Of these, 33 are in Sub-Saharan Africa, including Zambia, Senegal, Niger and Rwanda, but also Bolivia, Haiti, Honduras and Guyana in Latin America and Afghanistan in Asia. Some critics said that the initiative was designed more to protect those who lent the money than to help the people in the countries which, even at the end of the initiative, will have enormous and unsustainable debts.

Activities

1. Explore colonialism by writing a list of all the colonising countries and the countries that were colonised. Which colonisers had the most countries? Plot them on a map with different colours for each coloniser. Were there parts of the world that were not colonised? Compare your map to one which shows levels of development such as the HDI map – are there any similarities? Why might there be?
2. How true is it that the enormous debt repayments of poor countries are subsidising rich countries?
3. Do you agree with President Mkapa of Tanzania when he said, 'I encourage you in your advocacy for total debt cancellation because, frankly, it is a scandal that we are forced to choose between basic health and education for our people and repaying historic debt'? Why? Why not? What are the arguments on both sides? Which do you agree with more?
4. In what ways do LEDCs suffer by having to spend so much on debt repayments?

2 Sustainable solutions to the problem of unequal development

By the end of this section you will be able to:

- describe how Sustainable Development Goals aim to reduce the development gap.

Some background to the Sustainable Development Goals

Sustainable development means 'meeting the needs of the present, without compromising the ability of future generations to meet their own needs' (from G.H. Brundtland's *Our Common Future*, 1987).

Sustainable development aims to maximise the economic, environmental and social benefits of development while minimising the economic, environmental and social drawbacks. There is often conflict between economic development and the protection of the environment.

Traditionally, economic development was seen in terms of large-scale industrial projects such as dams, enormous factories, airports or major road developments. Sustainable development is more likely to focus on small-scale projects that are designed for the well-being of local people and have the minimum impact on the environment.

Sustainable development is designed to improve quality of life and standards of living.

This can be done by:

- using natural resources in a way that does not seriously damage the environment
- encouraging economic development that the country can afford and so avoid debt
- developing appropriate technology, that is technology suited to the skills, wealth and needs of the local people, which can be handed down to later generations (see page 167).

Eight **Millennium Development Goals** were agreed in 2000. These were to be achieved by 2015. They were:

- the eradication of poverty and hunger
- universal primary education
- the promotion of gender equality
- reduction of child mortality
- improvements to maternal health
- combatting some dangerous diseases such as HIV and malaria
- ensuring environmental sustainability
- developing a global partnership for development.

In the period 2000–2015, there were remarkable successes in some of these Millennium Development Goals:

- The number of people living on less than US$1.25 per day declined from 1.9 billion to just 836 million.
- Around 43 million fewer children of primary school age around the world were out of school.
- Around 90 per cent of countries in the world have more women in parliament than they had in 1995.
- The number of deaths of children under five more than halved to 6 million.
- The number of mothers dying in childbirth reduced from 380 for every 100,000 live births to 210.
- New HIV/AIDS infections fell from 3.5 million to 2.1 million.
- Piped drinking water was made available to 1.9 billion more people since 1990.
- Overseas development assistance from MEDCs to LEDCs increased by 66 per cent between 2000 and 2014.

Despite these successes, there are still many challenges and a new set of goals has been set: the Sustainable Development Goals (SDGs).

What are the goals?

There are 17 Sustainable Development Goals (SDGs), which together contain 169 targets that are aimed to be met by 2030. The goals were adopted by the United Nations in 2015. We will look at three of these in detail.

Figure 8 The UN's 17 SDGs.

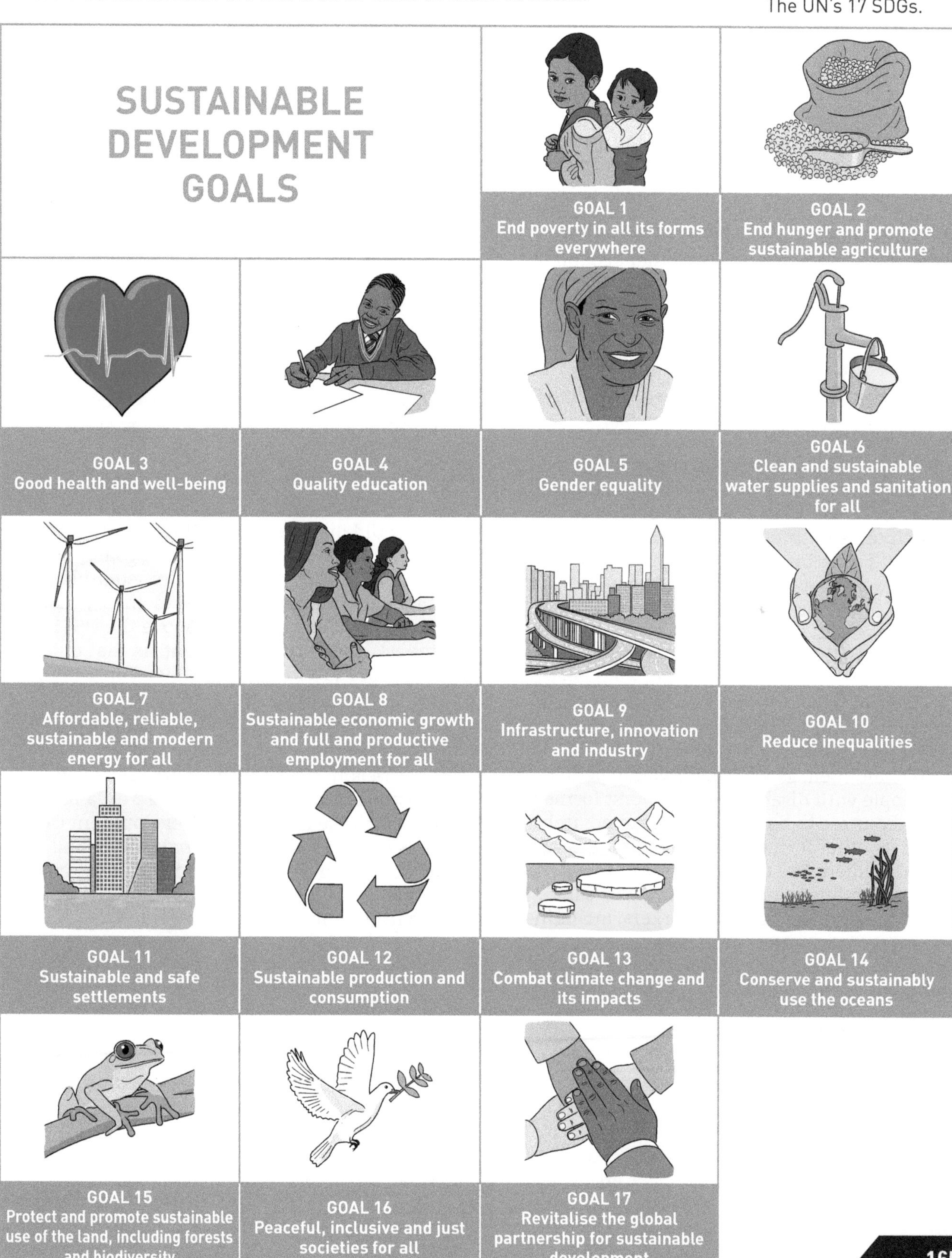

How does Goal 1 attempt to reduce the development gap?

While extreme poverty (defined as surviving with less than US$1.25 each day) has been reduced, one in five people in the world are still in that situation. Poverty is most harmful to children, who are often unable to access education and can have long-term health problems as a result of malnutrition in childhood. While wanting to eradicate extreme poverty by 2030, the goals also want to address less extreme poverty, to try to ensure that these people have access to basic services, land ownership, natural resources and microfinance. They plan to make them more resilient to climate change and other economic or environmental disasters. Policies in countries which are designed to help the poor will be encouraged to ensure that this section of the population is supported as much as possible.

How does Goal 8 attempt to reduce the development gap?

By 2030, the goal is to have sustainable growth of at least 7 per cent each year in everything produced (GDP) in the poorest countries in the world. It aims for more productive economies which will grow by diversification and the application of new technologies. Promoting creativity, innovation and entrepreneurship, and providing access to finance, the goal is to develop medium, small and micro economic enterprises. The plan is to achieve this without damage to the environment, by encouraging sustainable consumption and production. By 2030 there is to be full employment and decent work for all women and men, as well as for young people and people with disabilities. The worst forms of child labour are to be eliminated immediately, and all its other forms, including recruitment of child soldiers, are to be ended by 2025. The working conditions of workers must be safe and secure and the rights of workers, including migrant workers and especially women, are to be protected. Sustainable tourism is to be promoted to provide employment.

How does Goal 15 attempt to reduce the development gap?

By 2020 sustainable management of all types of forests will be promoted, deforestation halted and degraded forests restored. There will be more forests planted, with an agreed global target each year. Desertification will be combatted and the land and soil affected will be restored. By 2030, mountain ecosystems will be conserved to maximise the sustainable benefits that their biodiversity can provide to countries. By 2020, threatened species will be protected and saved from extinction. Urgent action will be taken to end poaching and smuggling of protected species of plants and animals. Ecosystems will be integrated with national and local planning, and consideration of biodiversity will impact upon development strategies to reduce poverty. By 2020, measures will be introduced to prevent the introduction of invasive alien species into ecosystems and significantly reduce their impact by controlling or eradicating the worst species.

Criticisms

There is little in any of these 17 goals that people could disagree about. However, there are criticisms. Some have said that the goals do not go far enough and that hunger, for example, should be eliminated in five years. In any case, they say, the lifting of people above the US$1.25 limit is not such an achievement, as that is not adequate to keep anyone alive. The limit of extreme poverty, they argue, is more like US$5.00, and this should be the target to raise people above.

Other criticisms are that there are too many goals, and they have been described as 'a mess' compared to the Millennium Development Goals. There is also a view that the goals are not all achievable – can you have rapid economic growth and yet combat climate change and not damage the environment at the same time?

Activities

1. Working with a partner, take any two of the Sustainable Development Goals not covered in this section. Take one each and make a note of the main points of the goals you have picked. Explain them to your partner and hear your partner's explanation.
2. Look at the criticisms of the SDGs. Are they fair? Are there any you agree with? How do you think they could have improved the SDGs? Do you think the goals will all be achieved? Does it matter if some are not achieved by 2030?

Sustainable development and appropriate technology

What is appropriate technology?

Appropriate technology, by its design, considers the community that it is intended for and the environment in which the community lives. Typically, appropriate technology uses fewer resources, is easier to repair, costs less to buy and to run and has a lower impact on the environment than other technologies.

A pump which requires fuel and, when it breaks down, needs expensive spare parts from abroad and specialised tools to fix it which local people do not have, is *not* appropriate technology. Something which uses locally available, fairly cheap materials and could be fixed by local people when it breaks *is* appropriate technology.

Has appropriate technology been successful?

One example of an appropriate technology product is the hippo water roller (https://www.hipporoller.org/). It was developed in 1991 by two South Africans to help rural women and children. It is generally women and children who have the responsibility of bringing safe drinkable water from wells and other water sources every day, a laborious and back-breaking task which can cause long-term neck and spinal damage. The water roller comprises a 90-litre drum with a steel handle to allow it to be easily moved, and it is a very efficient way to move water. Hygienic and durable, this technology requires no spare parts and has a lifespan of up to ten years. When it can no longer be used to transport water, the plastic barrel can be recycled into storage bins or used for washing clothes.

Evaluating the Hippo Water Roller

Reducing the time required to do the chore of water provision each day, children are able to spend more time at school and women are able to carry out other economic activities that they previously had little time for, such as selling agricultural produce. Since it was developed, it is said to be used in more than 20 countries across Africa, providing convenient water for up to 300,000 people.

Other appropriate technology products

A high-tech example of appropriate technology is Loband. This was the idea of two aid workers in Nepal who found it difficult to access the web in remote areas where there were slow internet connections. Loband works by simplifying webpages to allow them to download more quickly, allowing people to access the internet who previously got a very poor service. Aptivate, the non-governmental organisation (NGO) that invented Loband, now provides a range of IT technologies to communities in LEDCs, many of which are appropriate technologies.

You might wonder why appropriate technologies are not promoted more. For some time they have fallen out of fashion, being seen as a 'poor person's technology', and not promoted by governments or donors who wanted something more substantial to show that they were making a difference to communities. Few politicians would prefer to be photographed in front of a plastic drum or a composting toilet if they could be pictured instead in front of a dam or a new bridge.

By the end of this section you will be able to:

- define appropriate technology
- describe and evaluate the success of one product.

Research activities 

1. In a group, research three appropriate technologies and source a photograph or sketch of each.
2. Make a digital show of the images, and present to other groups the three alternative technologies you have chosen. Make sure that you explain why they are alternative technologies, and what their advantages are. If you have found any criticisms or disadvantages of the technologies, talk about those as well.

Figure 9 The hippo water roller in action.

By the end of this section you will:

- know and understand what fair trade is and the advantages it brings to LEDCs.

Fair trade

What is fair trade?

The aim of fair trade is for people in LEDCs who make or grow a product to get paid fairly for their work. It is about getting better prices as well as improved working conditions, local sustainability and fair terms of trade.

How does fair trade help LEDCs?

Producers in LEDCs get paid directly at fair prices, getting a fair share of the profits, which otherwise might mainly be concentrated in companies in the MEDCs. Fair trade sales really took off after 1988 when goods started to be labelled as fair trade, and sales continue to grow. There are now 4,500 products in the UK which are certified as Fairtrade and able to use the logo. Many are foods, such as coffee, sugar, honey, rice and spices, while non-food products such as beauty products, cut flowers and gold and silver are also able to be certified as having been traded fairly. Fair trade sales in the UK in 2012 were £1.57 billion. Almost one in three bananas sold in the UK is labelled as Fairtrade and, in 2013–14, 468,200 tonnes of bananas were sold on fair trade terms. The farmers and plantation workers received more than €19 million in Fairtrade premium payments, which are additional payments on top of the price of the bananas, for farmers and workers to invest in social, economic and environmental development projects.

Fair trade in coffee

Coffee is the most valuable tropical agricultural product traded in the world. However, coffee farmers only earn between 7 and 10 per cent of the price of coffee in the supermarkets in MEDCs. One-third of the price of coffee goes to the supermarket that sells it. Many coffee farmers receive market payments that are lower than the costs of coffee production, which keeps people in a cycle of poverty and debt. The world price for raw coffee Arabica, the Ethiopian plant now widely cultivated to provide our coffee, was US$3.88 per kilogramme in late 2016, which is up from US$3.26 since the beginning of that year, but much lower than US$5.46 in October 2011. When the prices are low, the people who produce the coffee have to work harder and longer, but may still get less money. The intensive production of coffee can cause other difficulties such as the loss of trees through deforestation to clear land for coffee cultivation and other environmental problems due to the use of pesticides.

Fair trade has helped coffee producers in a range of ways. Fair trade:

- guarantees a minimum wage for the harvests of small producers; this means that they can provide for the basic needs of their families and are much less vulnerable to poverty
- provides farmers with credit facilities and pays them a minimum price
- empowers communities to organise into co-operatives which helps their negotiating position; for example, to get higher prices for their coffee
- improves access to services for farmers such as training
- provides standards which encourage environmental protection
- encourages minimising or eradicating the use of toxic pesticides
- can provide access to finance to support farmers and improve the environment
- can support workers seeking to improve their rights and their terms and conditions of work

- brings the coffee directly to consumers and so cuts out some of the costs of middlemen, so that the benefits of the trade are more likely to reach the producers
- works with co-operatives so that the producers control the business and the members of the co-operative share the profits and benefits fairly.

Some specific advantages of fair trade for Kagera Co-operative Union, based in north-west Tanzania, have been:

- helping 60,000 smallholder coffee farmers to sell their coffee on the fair trade market
- providing training to grow healthy and productive crops
- improving facilities helping to build schools, mosques, clean water supplies and health centres, as well as bridges and roads to link the coffee farming areas to the farming communities
- helping to buy an instant coffee factory, the only one in Tanzania
- providing money to families to improve housing and promote education for their children
- promoting diversification into vanilla, potatoes and banana production, thus reducing the vulnerability of the agricultural communities
- providing opportunities for women to earn money independently.

How do MEDCs benefit from fair trade?

- If producers in LEDCs earn higher wages and are helped to develop in the long term, they will be able to increase their spending power. This will mean that they will be more able to afford to buy high-value manufactured products such as computers from the MEDCs. In turn this means that MEDCs will expand their trade and have new markets in which to sell their manufactured goods. This is becoming more important as the population, or number of potential consumers, of LEDCs rises, and that in MEDCs stabilises or even falls.
- Consumers in MEDCs will be able to buy top-quality products knowing that the trade has been good to everyone.
- Consumers in MEDCs are able to keep their consciences clear as they know that they have helped poor producers in LEDCs by trading in an ethical way. People will feel that they are being good citizens by caring for other people. Trading in a fair way will contribute to people's well-being and development in other countries, so improving the quality of life for everyone.

Is fair trade really fair trade?

It is certainly trade – £1.57 billion of Fairtrade-labelled products are bought in the UK alone – but is it really fair?

It has its critics. Some say that there is limited evidence that fair trade benefits producers. Even the best estimate of what proportion of the extra money paid by consumers gets to producers is 50 per cent. Others suggest a figure closer to zero. Starbucks sells Fairtrade coffee, but the company takes 55 per cent of what consumers pay, while only 10 per cent at most goes to those who grow it. Also, critics say that there is limited evidence to suggest that farmers actually get higher prices under fair trade. Some farmers cannot afford the fees to join the Fairtrade network, and remote and poorer farmers may not be able to organise together into co-operatives to pay that fee. The burden of getting Fairtrade certification falls on the producers. Some critics say that fair trade harms other farmers who are outside the fair trade arrangements.

Others criticise fair trade for not going far enough. By focusing on benefiting small groups of producers, they are not addressing the bigger issues of trade which are unfair. The fair trade movement works within the system of dominance of multinational corporations and mass retailers, rather than challenging this MEDC–controlled network of trade.

Those in favour of fair trade say that it addresses the injustices of conventional trade, and argue that Fairtrade works with big brands like Cadbury and Starbucks and also the big supermarkets to make fair trade mainstream and so helps more producers.

Luis from Ecuador

My neighbour sells his bananas through a middleman to the world market and he gets $1 for a 20 kg box of fruit. I sell my bananas to La Guelpa collective which is part of the fair trade market and I get $2.50 for every 20 kg box of fruit. This means I can have a much better standard of living.

The collective has invested its money into the village and farms. There is now clean water in all of the communities and we are trying to improve the health care and build more schools.

José from the Dominican Republic

My wife and I work a small farm, growing cocoa, and together with other farmers in the local farmers' co-operative (called CONACADO) we process, market and transport the cocoa. CONACADO sells the cocoa to UK fair trade chocolate producers, guaranteeing a better price for us all and enabling us to strengthen the co-operative. All my five children went to school and some to further training, although I cannot read or write.

Celestina from Tanzania

We get a higher price when we sell our coffee for 'Cafédirect' (a fair trade coffee). This means that our co-operative has been able to pay a doctor who will give treatment to our members.

(A co-operative is a group of people who have formed a business together. The members share the profits and benefits.)

The price difference has meant that I can afford more food for my family and send my children to school properly equipped with books for the first time.

Cafédirect stands by the agreed price for our coffee even when the international price of coffee falls on the world market.

▲ **Figure 10** Three stories of people who are benefiting from fair trade.

Activities 

1 You are trying to persuade your school or local authority to only buy fair trade products where possible. Prepare and present a lightning 1-minute presentation which presents the advantages of fair trade and gives the main reasons for your school or local authority changing to be fully fair trade. Do not use any props or digital media – you have to talk persuasively for one minute. Try not to read it out and remember to cover what is in it for *them* as well as any other benefits.

2 a Using the table below, draw a graph of the breakdown of the profits from a cup of coffee.

Who receives what percentage profits from a cup of coffee?	
Growers	10%
Exporters	10%
Shippers and roasters	55%
Retailers	25%

b Describe the main features of the graph.

c Who takes the greatest risks? Why do you think they take most risks?

3 Read the stories of Celestina, Luis and José (Figure 11). How do they benefit from fair trade?

4 Supermarkets are selling more and more Fairtrade products. Make a list of Fairtrade products sold in supermarkets. Sort the products into categories. Explain why the range of Fairtrade products in shops is increasing. Explain your answer from the point of view of:

i supermarkets

ii consumers.

3 Globalisation problems

By the end of this section you will:

- understand the meaning of the term globalisation
- be able to show how globalisation can help and hinder development in a BRICS country.

Defining globalisation

The world is shrinking. People can travel about much more quickly and easily across the planet. The internet means that everyone who can afford a computer or a mobile telephone can easily communicate within and between countries. World trade has opened up so that brands can be sold around the planet. This is called globalisation. Many industries have globalised as well, with branches in a great many countries around the world. These are called Multinational Corporations (MNCs).

Many MNCs are household names around the world: Coca-Cola, Nike, Google, McDonalds, Walt Disney, Facebook, Monsanto and Procter and Gamble are a few. Apple had a market value of US$724 billion in 2016. Others have names that are not household names but they are still enormously powerful around the world, affecting our lives in many ways. All of these companies are often richer than the countries in which they operate. Most of the market in major products is controlled by a few companies, as shown in Figure 11.

What does globalisation involve?

Globalisation means that:

- international operations are increasingly important for people and companies
- decisions taken in one country can quickly affect other countries because of the improvements in transportation and the spread of global communications, such as email and videoconferencing
- economic power is becoming increasingly concentrated in the hands of a few global companies or MNCs. It changes each year but in 2014, of the world's 100 largest economies, 37 were MNCs and not countries.

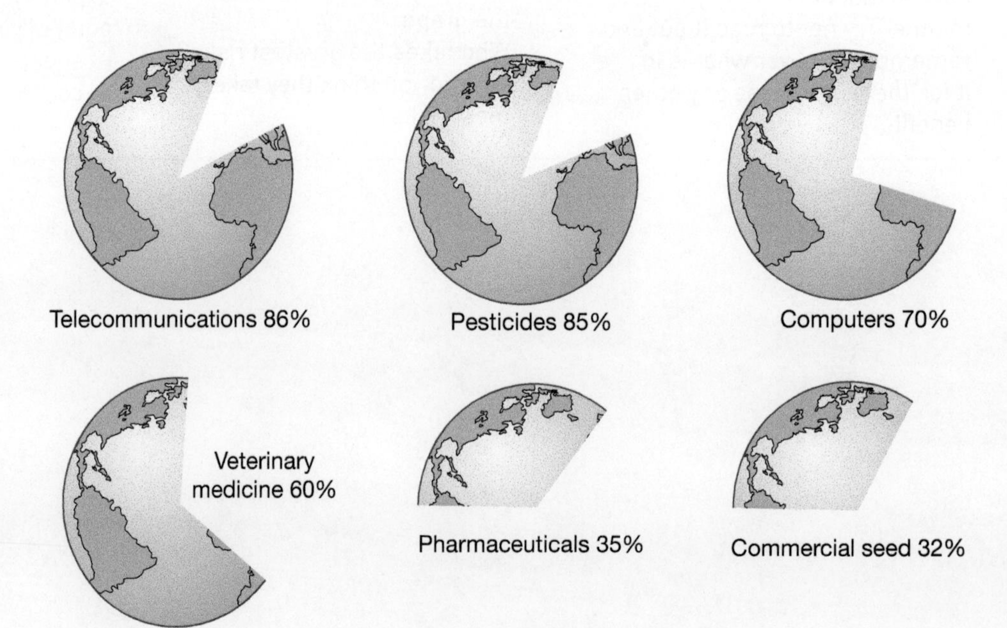

Figure 11 The market share of the top ten corporations by sector.

Wal-Mart, the US company that owns Asda in the UK, were 28th; Royal Dutch Shell, Exxon Mobil (Esso) and Sinopec-China Petroleum – the oil companies – were next, all above Austria, the United Arab Emirates and Denmark. Samsung, Volkswagen, Toyota, Apple, IBM and Nestlé also featured in the top 100

- countries are increasingly interdependent because they need to trade with one another and to exchange goods – the lives and actions of people in one country are increasingly linked with those of people thousands of kilometres away.

Globalisation affects some places in the world more than others. The headquarters of the MNCs are usually located in the MEDCs, and this is where most of the profits are likely to end up. On the other hand, LEDCs generally produce the raw materials. Sometimes the processing or assembly of the raw materials takes place in an LEDC to take advantage of the low wages there.

Another impact of globalisation has been the increased movement of people, goods, money and ideas, which is generally seen as a beneficial change. However, this movement has also led to the movement of invasive species and of diseases. The swine flu pandemic, which started in 2009, is an example of this. An example of an invasive species is *Obama nungera*. Originally from Brazilian rainforests, this flatworm has spread through soil in Dutch pot plants and is now found wild in the soils of Spain and France. It poses a threat to soils by reducing indigenous earthworm populations. In 2016, it was discovered in soil in England for the first time. On top of threatening the environment, invasive species such as New Zealand flatworms, zebra mussels and Japanese knotweed are thought to cost the UK economy around £1.7 billion each year.

Figure 12 Globalisation is thought by many to favour MEDCs.

1. **Global communications** – It is becoming cheaper and easier to communicate. Jake in New York can Skype his cousins in Cork for as long as he likes at no cost. Ten years ago, the same call would have cost US$3. In 1930, when Jake's grandfather first arrived in America from Cork, it would have cost him US$245 to phone home. If Kathy in the USA saves a month's wages, she can buy the most up-to-date computer. A worker in Bangladesh would have to save eight years' wages to buy a computer. Jake and Kathy can surf the web and understand four out of every five websites – because they are written in English. However, nine out of every ten people in the world don't understand English.

2. **Global advertising** – Advertisements for goods like Coca-Cola and Levis can be seen everywhere from Belfast to Berlin to Beijing. Nike reportedly spent US$3.031 billion in 2013–14 on 'Demand Creation', which is around US$100 per second. Demand creation is increasingly not just advertising but having sports stars wear the company's products. In 2013, Nike announced a sponsorship deal with Rory McIlroy, Northern Ireland's famous golfer. While the figures were not made public, it was widely reported that Nike would pay McIlroy up to US$20 million per year just to use its branded gear and sportswear, a deal extended for ten years in 2017. Workers in Nike's labour-intensive factories in Indonesia earn the equivalent of US$3.50 per day. Nike closed its factories in Japan, South Korea and Taiwan when demand for higher wages started to grow, relocating them to Indonesia, Vietnam, China and Thailand.

3. **Trade is increasing** – Three times more goods are shipped or flown between countries now than 20 years ago. Trade can create jobs in poor countries, but people often have to work in terrible conditions. On banana, coffee and flower farms, for example, workers get very low wages, work long hours and have to use dangerous pesticides or chemicals. Some countries are missing out on trade altogether: the 50 poorest countries are being left behind and making less and less money from trade compared to richer countries.

4. **Global rules** – The World Trade Organisation (WTO) was set up to write the rules for trade between countries and to make sure that all countries obey them. This could be very good – fair rules would mean that everyone would win. However, the rules are written to suit rich countries. This is because rich countries can afford to pay lawyers to debate the rules, while poor countries cannot. For example, Japan has 25 trade experts at the WTO, Bangladesh has one, and 29 of the poorest countries have none. There are up to ten meetings every day. Poor countries cannot go to all the meetings and so important decisions are made without them. Critics have argued that the focus is on trade and the creation of wealth, but that the environment and the rights of workers are ignored.

5. **Travel and tourism** – Travelling abroad for holidays can be interesting and exciting. People make friends from other countries and may discover many interesting things about other people's lives and the global environment. However, some people fear tourism can have a negative side. In many countries, local people dress in beautiful costumes and jewellery, such as the Maasai cattle farmers in Kenya. Sometimes, tourists treat local people like objects to be photographed rather than human beings who can feel and talk.

6. **The media** – Television, films, radio and newspapers can help people to learn about what is happening in other places and what it would be like to live there. On the other hand, the media may also give people a false impression of other countries. In India, many people think that life for everyone in Europe and the USA is exactly as it appears in the movies: rich and exciting. Meanwhile, news reports about Africa often only show images of desperate hungry people. However, most people in Africa have little money but find many ways to make their lives happy and full, and are certainly not helpless or starving.

▲ **Figure 13** Six situations in which globalisation affects people.

Globalisation is said to favour MEDCs at the expense of LEDCs. MNCs often have factories in many LEDC countries but the most profitable and highest-status parts of the business, such as research and development, are likely to be kept in the MEDCs. It is estimated that 86.5 per cent of MNCs are based in MEDCs; indeed, 56 per cent of them are in just one country, the United States. The UK is home to 16 per cent of them.

Having branches in many countries has allowed some multinationals to develop elaborate corporate structures to reduce or avoid paying tax in many of the countries in which they operate. Some LEDCs have objected to mining companies, for example, shifting profits out of their countries before they could be taxed. Zambia was losing an estimated US$2 billion each year through tax avoidance, largely by copper mining companies, which is more than the country spent on health and education combined. Mongolia's enormous copper mine was developed by a Canadian company which qualifies as Dutch for tax purposes as it has a branch there. According to Dutch law, money from the mine can be paid into the Dutch company, thus avoiding having to pay tax in Mongolia. Records show that there are no employees in the Dutch mining company, and it would seem to be a front to avoid paying tax, probably quite legally. Mongolia stands to lose US$5.5 billion over the life of the mine, almost as much as the GDP of the country. Both Zambia and Mongolia are fighting to change the tax rules.

Globalisation and BRICS

The acronym BRIC was coined in 2001 by a banker. He was referring to Brazil, Russia, India and China, which at that time were thought to be at a similar stage of economic growth, offering great potential in the increasingly globalised world. South Africa was added afterwards, making the abbreviation BRICS. These countries have large economies and were going through a period of rapid industrialisation in 2001, although Brazil's and Russia's GDP fell in 2015.

In 2015, these five countries accounted for half of the world's population, being home to 3.6 billion people. They produce 22 per cent of the world's GDP and their economies are predicted to grow rapidly, while many other countries are experiencing very little growth. The BRICS countries now have a summit every year when they explore ways in which they can benefit each other. They are considering setting up a 'New Development Bank' which will rival the World Bank and the International Monetary Fund. One of the reasons for that is that those two organisations are perceived to support the needs of MEDCs, and not countries such as the BRICS countries. They also plan to develop their information technology sectors, challenging the USA's traditional lead in this area. Many other countries including Afghanistan, Argentina, Indonesia and Turkey have indicated that they would like to join the group.

CASE STUDY

India in a globalised world

- India is a vast country with a population of 1.32 billion, just behind that of China. It has experienced rapid economic growth with a rise in GDP of 7.6 per cent in 2015. Even Ireland, the EU's fastest growing economy, only managed 3.5 per cent in 2015.
- India has been greatly influenced by globalisation. Since the 1990s the Indian economy has been opened up, barriers to trade have been demolished, entrepreneurs have been encouraged and investors from other countries welcomed. As a result, India's economy has grown at enormous rates. In 1996–97, growth rates of almost 78 per cent were claimed. This compares to a growth rate of just 3 per cent in the 1970s. The economy is still growing rapidly. It is now the world's third largest economy in terms of purchasing power and, despite the global slowdown, India's economy continues to expand.
- India is also recognised as global player as a member of the G20, a group of the world's 20 major economies. It is also hoping to land a permanent seat on the United Nations Security Council with some support from current permanent members.

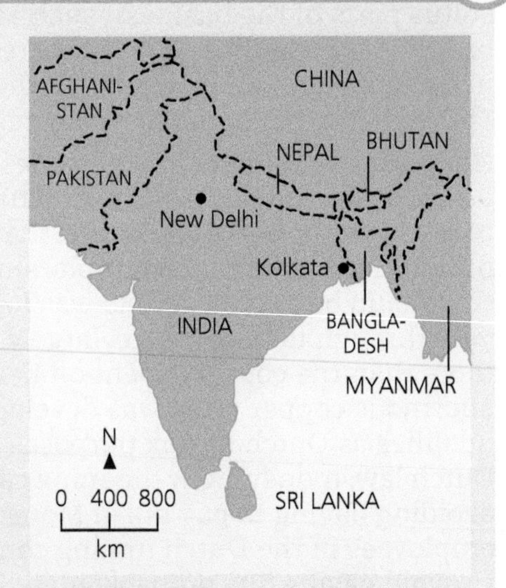

Figure 14 Map of India.

How has globalisation helped development in India?

- Supporters of globalisation believe that India's economic success has come about because of the opening up of the economy. India's low labour costs and its huge English-speaking workforce, combined with encouragement from MNCs, have made the country a popular destination to locate everything from manufacturing to call centres.
- The success of high-tech industries in India has resulted in the return of large numbers of skilled Indians who had moved overseas to get work. This has been described as a 'brain-gain', the opposite of 'brain-drain', which is when people leave. These people form part of India's growing middle class, a potentially vast domestic market for India's industry.
- Average living standards appear to be increasing in India as a result of globalisation. Between 1990 and 2015, life expectancy rose from 59 to 68 years. Between 1990 and 2011 adult literacy grew from 50 per cent to 74 per cent.
- Mass consumption, once only found in MEDCs, is now available to more and more Indians. Enormous shopping centres are being built in all the major cities and towns to allow the middle classes to spend their money. Prestige office blocks have been built in some of the cities, such as the Infosys headquarters in Bangalore (see Figure 15). Hyderabad, at the heart of high-tech India, is reputed to be planning to build the tallest building in the world.
- It has been suggested that, if economic growth continues, this will drive up living standards for the whole population of India, and globalisation will be identified as the cause.

Figure 15 Infosys headquarters in Bangalore.

Another view of globalisation

- There is an alternative view. Despite globalisation and the claims made of it, 455 million Indians were forced to try to survive on less than US$1.25 per day in 2005, an increase of 35 million people since 1981. The World Bank reports that nearly 58 per cent of India's population earned less than US$3.10 per day. Almost half of children under 5 years of age are malnourished, and in fact more than one-third of all malnourished children in the world live in India. By contrast, one-third of the children from the wealthiest families in India are over-nourished, and childhood obesity in India is estimated at almost 13 per cent. Less than one-third of homes in India have a toilet. Of India's 500,000 villages, less than half have a connection to the electricity network. However, around 156,000 Indians were millionaires in 2013, and that is forecast to rise to almost 360,000 by 2018.

Figure 16 A shopping mall in Mumbai.

Percentage (number) of Indians who are middle class	20–25% (300 million people)
Increased spending by India's middle classes	US$300 million
Numbers of mobile phones in India	20 million in 2003 350 million in 2008
The number of Indians with the same spending power as US residents	40 million
Percentage of India's population younger than 40 years of age	40%
The increase in Indian 'super-rich', with assets worth more than £3 million, in the next five years	Numbers will triple to 220,000

Figure 17 Some facts about India from a range of sources.

- Globalisation has been accused of widening the gap between rich and poor. Opening up trade between India and other countries has benefited those who have the skills to work in the high-tech industries and in the call centres, but has harmed the chances of ordinary Indians as goods have come in from outside, damaging their chances of getting a job.
- Some have blamed the opening up of trade by globalisation as something which has started to erode India's traditional values, putting consumerism above the family and community which were once so important in India. Globalisation has brought Western ideas of dress, diet and behaviour, some of which is shocking to older or more traditional Indians.

Figure 18 A homeless man begs outside a car showroom in Kolkata. Globalisation has done little to reduce poverty in India.

The new shopping centres may be crowded but they are mocked by some as a place to buy scented candles and other tasteless articles by people with nothing else to do with their money.

- India is a richer country overall because of globalisation, but the benefits of this wealth have been too slow in helping India's poor to have a better life. Average income per person per year was just US$616 (£500) in 2013, the lowest of all the BRICS countries. This is an average figure and there are a growing number of millionaires and billionaires in India. Many of the poor in remote rural areas, far away from the rapidly growing cities of Mumbai, Bangalore or Hyderabad, have to try to survive on much less.
- This is not just a problem for all those millions with low incomes. It could threaten India itself. As the gap between rich and poor widens, there has been widespread unrest from communist militants across many areas of India. According to India's Prime Minister in 2006, they are 'the single biggest internal security challenge for India'.
- The communist guerrillas are not likely to beat India's government and its huge army, but they can still cause problems in some of the poorest regions of India, where they are strongest. The areas may continue to get little investment and so will not benefit from the economic growth happening elsewhere. This will further increase the differences between the richer and the poorer parts of India, those places which have benefited from globalisation and those which have not. While globalisation did not cause the rise of the guerrillas – they started in the 1960s – the fact that they are still there is because of local support, and the increased gap between rich and poor in different parts of India is likely to provide continued support for these communist guerrillas.
- It is unclear where globalisation will take India, and indeed the rest of the world. Continued trade deals open up more markets and will allow more wealth to be created. If that wealth is in the hands of a few, and does not benefit everyone in society, increasingly questions will be asked about how good a thing globalisation has been.

Activities

1. Research a BRICS country. Choose either Brazil, Russia, China or South Africa.
   a. For your chosen country, research the country's economic growth since 2000. How much richer is the country now than in 2000? Is there evidence that the growth is continuing, or does it seem to be slowing down?
   b. Find some images which illustrate how the wealth has changed the country – this might be large infrastructure projects or new buildings.
   c. Look for evidence of whether this increase in wealth has benefited all the people of the country, or just a few.
2. In groups, prepare a digital presentation on Globalisation. Each slide should have text – not too much, and not 'copied and pasted' – and an appropriate image. Decide beforehand what the success criteria will be. You might want to base that on this list:
   - a slide giving a definition of globalisation
   - two slides on the benefits of globalisation
   - two slides on the disadvantages of globalisation
   - a final slide showing the impact of globalisation on you
   - finish with an image which summarises globalisation for your group.
3. In groups, gather a range of photographs relating to globalisation. You might use a search engine and search under 'globalisation' and select 'images', rather than web or videos or news, etc. Note that American English will spell it 'globalization', so you might want to try searching under that term too.
   a. Pick the first 20 or so images, skipping those that seem to be similar. Try to get some with people in them.
   b. Choose one from those that you have downloaded that make you uncomfortable, sad or angry, and explain to the group why you chose that one.
   c. Each group could choose a photograph and place it in the centre of a digital presentation. Ask questions like this about it:
      - who created the image?
      - why did they create it?
      - who is involved?
      - what is its message?
      - is it 'for' or 'against' globalisation?
   d. Record your questions and the answers of your group in text boxes around the image. Share your image and its questions with other groups and read theirs.

Sample examination questions

1 Social and economic indicators are often used to measure development. State one economic indicator of development. [1]

2 Study Figure 1, which shows the Human Development Index (HDI) for countries in South America. Answer the questions that follow.

(i) Using Figure 1, describe the variations in Human Development Index in South America. [4]

Figure 1 Human Development Index in South America, 2015.

(ii) What are the advantages of using the Human Development Index over other measures of development? [4]

3 (i) Circle one of the following which would be appropriate technology for rural areas in LEDCs? [1]

Airport
Clean water supply
Mobile phone masts
A supply of tractors made in MEDCs

(ii) Describe and evaluate one appropriate technology product. [5]

4 Globalisation has sometimes been blamed as benefiting the rich and privileged, but leaving many others behind. Using a case study of a BRICS country, discuss how globalisation can hinder development. [4]

5 (i) State the meaning of the term 'fair trade'. [2]

(ii) Explain one advantage that fair trade can bring to an LEDC. [4]

UNIT 2

LIVING IN OUR WORLD

THEME D: Managing Our Environment

▲ A photovoltaic solar power station in Morocco.

What are the advantages of solar power for countries and for the environment?

By the end of this section you will be able to:

- describe the greenhouse effect and define carbon footprint
- understand how both of these contribute to climate change.

Contributions to climate change

Average temperatures across the world are rising. Scientists increasingly agree that, if they were allowed to rise by 2°C, the effects would be serious. They would include severe droughts, more destructive storms and flooding and widespread food and water shortages. Some even say that we risk making the planet uninhabitable for future generations.

The rise in temperature is an effect of climate change. It means polar ice caps melt even faster than they are melting at present, causing a rise in sea levels which will drown coastal regions, including many of the largest cities in the world. Changes in the climate may also cause the extinction of many plant and animal species, leading to falling crop yields in some parts of the world and encouraging the spread of diseases.

What is the greenhouse effect?

Year	Average rainfall (mm)
2000	1337.3
2012	1330.7
1954	1309.1
2008	1295.0
2002	1283.7

Figure 1 The five wettest years in the UK since records began in 1914 (Source: Met Office).

Of the first 14 years since 2000, 13 were the warmest on record. 2001 to 2010 is the hottest decade on record. December 2015 was the wettest month ever recorded in the UK, at the time of writing – these records keep getting broken. In the UK, four of the five wettest years ever have occurred since 2000.

Each of these years brought floods to many areas of the UK and Ireland, destroying people's possessions and ruining their homes and livelihoods. These changes are much more important than just for our country – this is a global problem. World temperatures are estimated to have risen by 0.5°C in the twentieth century and could rise by up to 5.8°C by the end of the twenty-first century (see Figure 2). This has been called global warming but the term is a bit misleading as not all places will see temperatures rise at the same rate, even though the average temperature of the planet is increasing overall. A more accurate term is climate change. This emphasises that it is not just higher temperatures but changing precipitation patterns and wind speeds that will change. More extreme events – floods, drought and storms, for example – will become more frequent all around the world. These will impact on people in a range of ways, from creating dangerous conditions for many people to destroying crops and homes and disrupting communications. Climate change also threatens ecosystems on which we depend.

The mechanism that creates climate change is the greenhouse effect (see

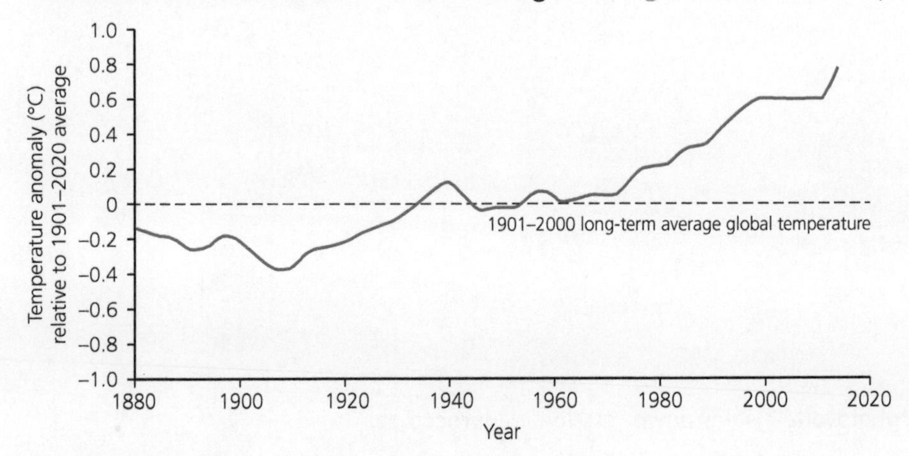

Figure 2 Average global temperatures, 1880–2015.

Figure 3). The energy from the sun passes through the atmosphere and heats the earth. That heat radiates out into the atmosphere but, as it is long wave radiation, some of it is trapped by gases in the atmosphere. It is this greenhouse effect which makes the Earth able to sustain life. However, people are adding further greenhouse gases to the atmosphere, through activities like burning fossil fuels and clearing forests. This causes more heat to be trapped in the atmosphere, which can change the climate. Climate change is caused by an increase in the amount of greenhouse gases which trap heat in the atmosphere; the main greenhouse gases are carbon dioxide, nitrogen dioxide and methane.

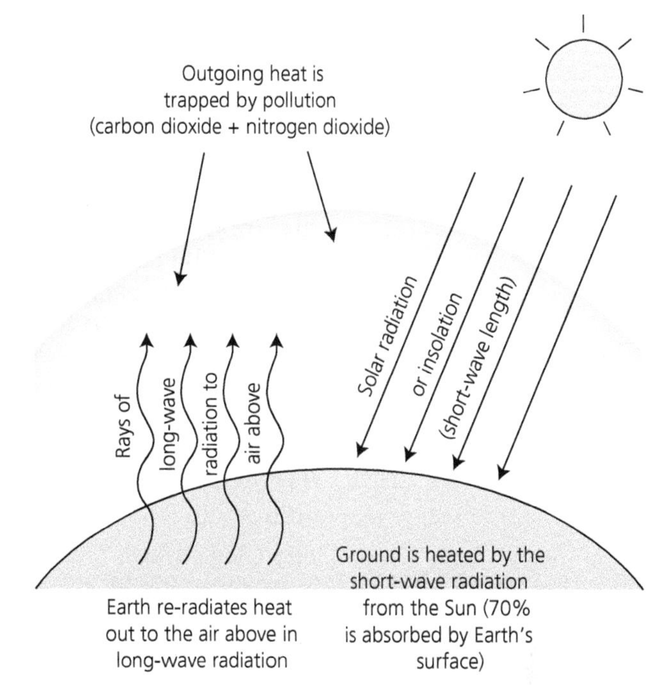

Figure 3 The greenhouse effect.

What is meant by 'carbon footprint'?

Using *any* resource has an impact on the environment. If we eat a can of beans there is a cost in growing the beans: natural vegetation removed to create the farmland, water used to irrigate the beans, chemicals used to reduce pests, power used to plough the soil and harvest the crop, power used in the processing, the mining to produce the steel and tin used in the metal containers and the energy used to make those,

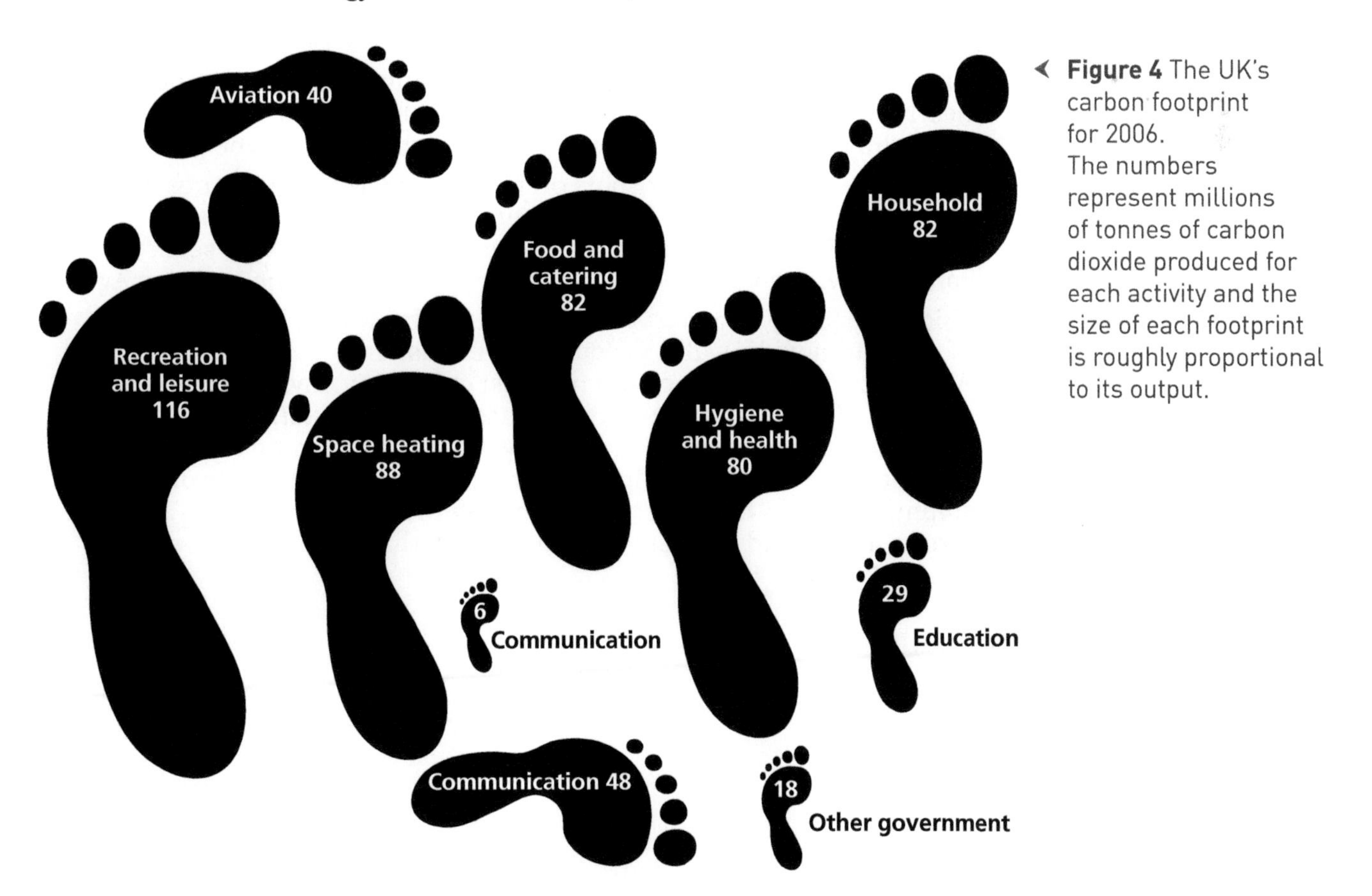

Figure 4 The UK's carbon footprint for 2006. The numbers represent millions of tonnes of carbon dioxide produced for each activity and the size of each footprint is roughly proportional to its output.

the transporting of the beans and the tins to the processing plant and the fuel used to transport those to your local supermarket. Add on the lights in the supermarket and the fuel you use to go to the shop and you have an idea of the total cost of using even the most basic of resources.

One of the best ways of working out how costly using a resource is to the environment is by working out the carbon footprint. This is a measure of the impact our activities have on the environment, particularly on climate change. The carbon footprint is a measurement of all the greenhouse gases that we each produce converted into the equivalent weight of carbon dioxide produced so that they can be compared. The idea of a carbon footprint is useful as consumers can easily grasp the impact of their activities. However, calculating the carbon footprint is very difficult. For example, making a product outside the UK generates carbon, and if a consumer in the UK buys that product, that adds to their carbon footprint. It is challenging to know how much to add on, as the way the product is manufactured might not be clear.

The UK's carbon footprint for 2006 is shown in Figure 4. The carbon footprint peaked in 2007, at the equivalent of 1,296 million tonnes of carbon dioxide. By 2013, it had dropped by 19 per cent from that peak. However, the drop was not constant, and the UK's carbon footprint actually rose again, although just by 3 per cent, between 2012 and 2013.

How do the greenhouse effect and our carbon footprint contribute to climate change?

Both human activity and natural causes contribute to climate change.

Natural causes of climate change

The following are examples of natural causes.

- Natural climatic cycles. The Earth's orbit changes very slightly between nearly circular and more elongated every 100,000 years. This cycle is evident in the glacial/interglacial cycles of roughly the same period.
- Sunspots – slightly darker spots on the sun where temperatures are temporarily lower – can also affect temperature. The period between 1645 and 1715, a time during which very few sunspots were seen, coincides with a very cold period in Europe known as the Little Ice Age.
- Volcanic activity. When a volcano erupts it throws out large volumes of sulphur dioxide, water vapour dust and ash into the atmosphere. Although the volcanic activity may last only a few days, the large volumes of gases and ash can influence climatic patterns for years. The gases and dust particles partially block the incoming rays of the Sun, leading to cooling. Sulphur dioxide forms small droplets of sulphuric acid in the upper atmosphere which reflect sunlight, and screen the ground from some of the energy that it would ordinarily receive from the Sun.

Human causes of climate change

Carbon dioxide is responsible for 50 per cent of climate change. It is feared that as rainfall belts shift, areas now covered by tropical rainforest could change into grassland or even desert. These would accelerate the rates of warming as fewer trees means an increase in the amount of carbon dioxide in the atmosphere.

There are three main ways in which human activities are major sources of greenhouse gases:

- Burning fossil fuels. In the past 200 years the need for more energy has grown as industrial development, population growth and prosperity have all increased. Most of this energy has come from burning fossil fuels. When coal, oil or gas is burned in power stations to generate electricity, gases such as carbon dioxide are emitted into the atmosphere adding to the greenhouse gases in the atmosphere.
- Vehicles. During the twentieth century more and more vehicles were using the roads, a trend that is continuing into the twenty-first century. The exhausts of cars and lorries emit polluting gases such as nitrogen dioxide, which add to the greenhouse gases in the atmosphere.
- Agriculture. Feeding the world's growing population requires more animals to be raised in those areas of the world where meat eating is common. Beef in particular has a high environmental impact and it is estimated that 15 per cent of all greenhouse gas emissions comes from agriculture. Methane production from large numbers of cattle is increasing very fast and so is nitrogen dioxide given off by fertilisers. Some scientists have argued that the biggest change that people could do to reduce their carbon footprint is not to stop driving their cars, but to eat less beef.

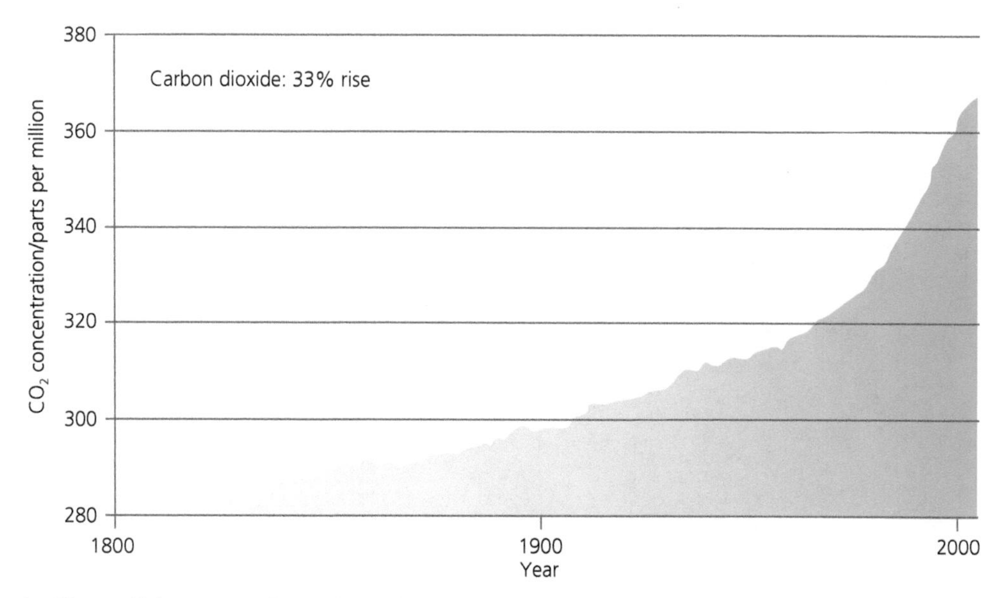

▲ **Figure 5** Increases in carbon dioxide levels since 1800.

Reducing our carbon footprint – that is, the amount of carbon emissions in the world – will help slow down climate change.

Press release from NASA, 20 January 2016

Earth's 2015 surface temperatures were the warmest since modern record keeping began in 1880, according to independent analyses by NASA and the National Oceanic and Atmospheric Administration (NOAA).

Globally averaged temperatures in 2015 shattered the previous mark set in 2014 by 0.13 degrees Celsius. Only once before, in 1998, has the new record been greater than the old record by this much.

The 2015 temperatures continue a long-term warming trend. NOAA scientists agree that 2015 was the warmest year on record, with 94% certainty.

"Climate change is the challenge of our generation," said NASA Administrator Charles Bolden. "It is a key data point that should make policy makers stand up and take notice – now is the time to act on climate."

The planet's average surface temperature has risen about 1.0 degree Celsius since the late 19th century, a change largely driven by increased carbon dioxide and other human-made emissions into the atmosphere.

Figure 6 NASA press release, adapted, 20 January 2016.

Figure 7 Ice sculptures by Brazilian artist Nele Azevedo melt on the steps of Berlin's Concert Hall at the Gendarmenmarkt on 2 September 2009. The event, which saw participants place some 1,000 ice sculptures, was sponsored by the WWF to attract attention to the earth's melting poles due to global warming.

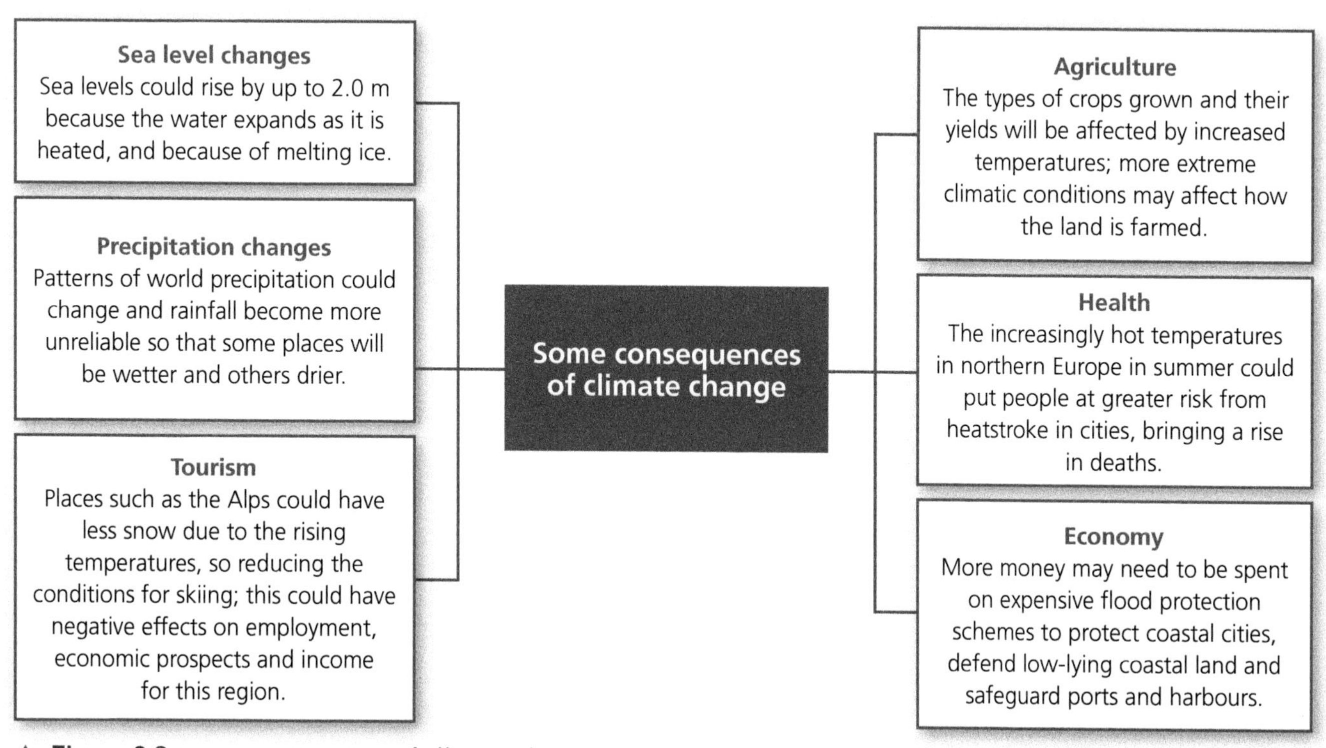

Figure 8 Some consequences of climate change.

Research activities

1. It is widely agreed that people are behind climate change and it is what we do that is increasing greenhouse gases and changing the climate. Find three images that illustrate the three main ways in which people cause climate change. Around each image write how these activities are contributing to climate change. In a different colour, list things that might be done to reduce the impact.
2. We know about the amount of carbon dioxide in the atmosphere in the past from direct measurements since around 1960. To go back before that, we can drill into ice sheets, especially in Antarctica, and take the measurements from the ice. Do some research into how and where this is done, and make a three-slide digital presentation of it. Include at least one map and one photograph.

Activities

1. There are a lot of figures and dates in this section. Draw a timeline and put as many of the facts and dates onto it as possible. You might also want to put key world events onto it, to provide a background to the climate change data.
2. Use one of the online carbon footprint calculators to check your own carbon footprint. You could find one by putting 'What is my carbon footprint?' into a search engine.
3. What does the result tell you? Is there any way that it might be inaccurate? Is it a useful way of looking at a person's impact on the Earth?

By the end of this section you will be able to:

- evaluate the effects of climate change on the environment, people and the economy.

The effects of climate change

What is the effect on the environment?

Climate change is happening at an extraordinary scale and pace, placing great stress on ecosystems. Each species adapts in response to climate change. Those changes impact on other species which in turn will impact on others, potentially causing a cascade of impacts throughout the ecosystem.

Extinction risks

So far the number of species lost as a consequence of climate change has been small, but it is forecast that, if the temperature were to rise by 2 or 3°C (which is on the low side of the projected changes), 20–30 per cent of all species around the world could be extinct in the next hundred years. This could potentially mean between 300,000 and 600,000 species being lost forever, with a devastating impact on ecosystems.

Coral bleaching

Figure 9
Coral bleaching.

Oceans are becoming more acidic as they pick up more of the increase in carbon dioxide from the atmosphere. This increase in ocean acidification combined with rising water temperatures is damaging coral reefs. Coral reefs are enormous underwater structures built by tiny animals which come together in large numbers in shallow tropical waters. These are rich ecosystems in their own right, and the role they play in protecting coasts and in fisheries and tourism means they have enormous global economic value. When sea temperatures are higher than normal over a long period, coral reefs suffer 'bleaching'. This means that they lose the colourful algae which the corals rely upon and which rely on them. What follows is a loss of colour in the coral reefs, greater exposure to disease and often the death of the coral.

When bleaching occurs on a global scale, it is referred to as a 'mass bleaching event', and there have been six of these since 1979. One mass bleaching event in 1998 killed almost 16 per cent of the world's reef building corals. Back to back `severe' bleaching of the Great Barrier Reef, a UNESCO World Heritage site in Australia, devastated it for 1,500 kilometres (900 miles). This is two thirds of the whole reef, and there are fears that the reef may never fully recover. Even the most optimistic climate projections would lead to the bleaching of 80–100 per cent of the world's corals by 2080.

Changes in seasonal life cycle events

These are seen in many ways, such as changes in bird migrations, butterflies emerging from cocoons before they did in the past, or plants flowering earlier than they used to. However, these changes in seasonal events may not be synchronised. For example, European Pied Flycatchers migrate from their Western African winter feeding grounds to European countries, including parts of the UK and Ireland, to breed each summer. The number of these birds fell by over 50 per cent between 1995 and 2014. Scientists believe that the cause may be that the breeding is out of sequence with the peak availability of food, as warmer springs lead to small insects and caterpillars emerging before the birds arrive from their migration. The birds need these as food for their young. The breeding now misses peak food availability, so fewer chicks are surviving and the bird numbers are falling.

Disruption to ecosystems

Ecosystems are highly vulnerable to climate change, particularly when combined with other pressures such as pollution, loss of habitats and invasive non-native species. While some species might benefit from climate change, most are likely to lose out.

In the UK, as elsewhere, many species will try to adapt by moving northwards or further up into mountains as average temperatures rise. Habitats will come under increasing pressure. Salt marshes will be threatened by rises in sea levels while some broad-leafed woodlands will suffer from increases in summer droughts.

Loss of Arctic ice

Recent research has shown that the average person from an MEDC produces enough carbon emissions to destroy 30 square metres of Arctic sea ice every year. It is forecast that, within about 20 years, the Arctic will lose all of its sea ice in summer. By losing sea ice, we lose the white polar cap which previously reflected sunlight out into space. This will speed up melting in the Arctic, and the Greenland Ice Cap will melt much more rapidly. Scientists also link the melting of the Arctic sea ice to changes in the upper-level atmosphere, which in turn is linked to more extreme weather events such as droughts in California and floods in western Europe. We are also losing a very beautiful landscape which has been inhabited by native Inuit people for thousands of years and is home to many native species such as polar bears.

	YEARS			Level of confidence that this change will happen
	2020s	2050s	2080s	
Opportunities				Low High
Less cold winters so fewer winter deaths				
Improved wheat yields because of warmer weather				
Opening of Arctic shipping routes as polar ice caps melt				
Reduced need for home heating				
More tourism in the UK				
Faster Sitka Spruce tree growth in Scotland				
Ability to grow new crops not possible before				
Threats				
More damage to homes because of flooding				
More pests in forests				
Increased energy demand for cooling buildings				
Changes in species migration patterns				
Shortage of public water supply				
Spread northwards of invasive non-native species				
Increased soil erosion because of heavy rainfall				
Increased need for irrigation water for crops				
Decline in productivity of some fish and shellfish				
Increased deaths due to higher summer temperatures				
Lower summer river flows				
Flood risk to high quality agricultural land				
Increased frequency of sewers overflowing				
Agricultural land lost because of coastal erosion				

KEY

Positive consequences	
	High
	Medium
	Low

Negative consequences	
	Low
	Medium
	High

▲ **Figure 10** Climate change risk in the UK (Source: Adapted from the UK Climate Change Risk Assessment 2012 Evidence Report).

What is the effect on people?

There may be some benefits for some people as a result of climate change. However, many more will suffer.

Water around the world

Both the quality and supply of fresh water are threatened by climate change. Rising temperatures are melting glaciers at an unprecedented rate. Glaciers in the Himalayas and other mountain ranges are shrinking, and many are forecast to disappear before the end of this century. Additionally:

- It is forecast that more precipitation over mountains will fall as rain rather than snow. While this might sound like a solution to water shortage, this rain will mostly fall in winter and flow off the land quickly, perhaps causing damaging floods.
- In the summer droughts, there will be little rainfall and no snowmelt to provide fresh water supplies.
- Billions of people depend on water supplies from the rivers which flow all year round from the world's mountains. For example, the Yangtze, the Brahmaputra and the Ganges all originate from Himalayan glaciers and sustain enormous populations in South East Asia – around 2.4 billion people live in the drainage basin of the Himalayan rivers. The Ganges alone provides water for drinking and agriculture for more than 500 million people. In North America, the water from glaciers in the Rocky Mountains and Sierra Nevada will also be affected. All of these places will have water supplies that are uncertain, and so water security is at risk.

Water security in the UK

The UK could also face its own threats to water security. Declining summer river flows and increased evaporation will contribute to water loss which could result in water shortages and restrictions on use. The UK Government estimates that, by the 2050s, 27–59 million people may be living in areas which are short of water. This does not take into account increasing populations.

Flooding

The UK Government estimates that annual damage from flooding could increase to up to £12 billion in England and Wales alone by the 2080s, from current costs of just over £1 billion. Areas vulnerable to flooding often have critical infrastructure such as electrical generation plants, schools, hospitals, water treatment plants and roads and railways. Many houses are threatened, with 330,000 currently at risk from floods, though climate change could increase this to between 630,000 and 1.2 million properties by the 2080s. This will have an economic as well as a social cost.

Drier summers

Climate models suggest that the UK could have warmer and drier summers in the future. This is one of a number of benefits (see Figure 10 on page 189), but there are likely to be more droughts and extreme events such as heatwaves. These will put considerable pressures on health care services, particularly for older people, and will likely disrupt transport, for instance through the buckling of railway lines.

Figure 11 A flooded road in Lisburn, Northern Ireland.

Fishing and aquaculture

The livelihoods of 520 million people depend on fish production, many of them small-scale fisheries in LEDCs. One-third of the world's population rely on fish and other aquatic products for the protein in their diet. Both aquaculture and fishing are threatened by climate change as rising sea temperatures and increased acidification of the oceans impact on the ability of fish to reproduce. This has been made worse by overfishing. Ecologists and economists have warned that worldwide fisheries could totally collapse by 2048, if no action is taken to stop current levels of overfishing. If countries keep fishing at their current rates, there is a danger of fish stocks declining to such an extent that it would make commercial fishing impossible, and recovery unlikely.

Farming

Depending on the assumptions used in the climate change models, agriculture in some places will be helped by climate change. Benefits are mostly predicted in areas with higher latitudes (which by chance are where MEDCs are mostly located), yet there will be strong negative impacts for populations poorly connected to trading systems, which includes many African countries. In these places it is likely that food production will fall and the traditional crops will not produce as much as they have before.

One study suggested that, by 2030, southern Africa's maize production, on which most farms depend, could fall by one-third due to climate change. In South Asia, yields of rice, millet and maize could fall by 10 per cent. The average crop yield in Pakistan is expected to be halved. On the other hand, maize production in Europe is expected to grow by up to 25 per cent.

Wider social problems

The impact of climate change on the supply of raw materials or food could also impact upon the UK and Ireland in other ways. It might lead to international instability. The UK and Ireland could experience effects ranging from changing food prices and availability, to potentially increased pressure of migration from environmental refugees fleeing droughts, floods and famines.

What is the effect on the economy?

As sea levels rise, weather patterns shift and extreme weather events become more common, there will be an impact on food production, water resources, ecosystems and human health. These will occur over many generations and affect large swathes of the Earth's surface. While it is very difficult to predict the exact economic costs of climate change, there are a number of studies which give us some idea.

The economist Nicholas Stern published a report in 2006 on the cost of climate change. The Stern Report concluded that, without action on climate change, the overall costs would be equivalent to losing at least 5 per cent of global gross domestic product (GDP) each year, an impact which would last forever. That would be the equivalent of the world being US$3.88 trillion poorer (a trillion is a million million). If a wider range of risks and impacts were included, this could increase to 20 per cent of global GDP or more being lost, again forever, which is the equivalent of losing US$15,500,000,000,000 of the world's wealth.

More recent work suggests that Stern may have underestimated the cost of not acting on climate change. Scientists in Stanford University in the USA have produced a map showing the potential winners and losers from climate change, if no action is taken. Their research indicates that:

- India is forecast to have a 92 per cent reduction of GDP by 2100 due to climate change.
- Brazil's economy is expected to fall by 83 per cent and Saudi Arabia's by 96 per cent in the same period.
- There are a few countries that are forecast to grow, and both the UK and Ireland are predicted to have their economies grow by over 40 per cent in the same period, if no action is taken on climate change.

However, as not taking action would have a disastrous economic impact on most of the world, such gains would hardly compensate. Many countries are now beginning to consider that combatting climate change may actually make good economic sense. Denmark has an ambitious target to cut greenhouse gas emissions by 40 per cent by 2020. The former Danish Prime Minister, Helle Thorning-Schmidt, said that this would not negatively impact on her country's economy. 'There is no reason to think that [tackling] climate change isn't good economics ... we continue to grow and create jobs', she said.

The former Irish President, Mary Robinson, appointed as UN Special Envoy for Climate Change in 2014, argues that businesses have to change to become ever more environmentally aware, saying that sustainability has to become a core part of how they work.

The case of China

China has had remarkable growth in the last 25 years, with growth rates of 15 per cent in some years. Some have said their attitude was 'pollute first, and clean up later'. However, one of the most important initiatives in global climate change has been China's decision to speed up its move to cleaner energy production, and to confront climate change directly. China's leadership in the fight against climate change was vital in the Paris Agreement (see pages 202–203) and sent a strong signal that action was needed. They increasingly see climate change not just as a threat, if not tackled, but also as an economic opportunity.

In 2013, there was heavy smog in Beijing and other parts of northern China, affecting 600 million people. Beijing has now decided to close its four coal-fired power stations. Across China, the aim is to produce 20 per cent of energy from non-fossil fuels by 2030. Tackling air pollution in China will improve the health of the Chinese people. It will also ensure more modern infrastructure and make the country more economically competitive. Reducing pollution will bring investment, create employment, improve living standards and extend life expectancy.

China's GDP per person is currently US$7,500. China's emissions are expected to peak when GDP reaches around US$14,000 per person. This is the equivalent of each person emitting 8.2 tonnes of CO_2, mainly through the use of fossil fuels. While this sounds high, at their peak, the United States, Australia and Canada had emissions of 20 tonnes of CO_2 per person and Japan and the European Union peaked at around 10 tonnes per person. After reaching this peak, China's emissions per person will fall.

The changes in China will produce:

- better health in the population
- a reduction in medical costs.
- more jobs:
  - in renewable energy
  - in transport and construction.
- new jobs:
  - in energy conservation and environmental protection
  - in energy-efficient transport
  - in producing more environmentally friendly energy.

Activities

1 Work with a partner to produce a mind map showing the impact of climate change on the environment, people and the economy. Include figures where possible, and small diagrams where they would help someone to understand.

2 Use Figure 10 (page 189) showing climate change risk in the UK to identify the opportunities and threats that climate change might bring, based on medium emissions.

a Prepare a digital presentation with three slides to show some of the opportunities, with photographs to illustrate them where possible.

b For another three slides show some of the key threats, again using illustrations where possible.

c For each, note the timing of these effects, and how confident scientists are that they will happen.

d In a final slide, give your opinion about what is likely to happen and what the most significant changes will be in your lifetime.

▲ **Figure 12** Smog in China.

By the end of this section you will be able to:

- describe the waste hierarchy and how to 'reduce, reuse and recycle'.

Managing resources

What is the waste hierarchy?

The UK produces around 200 million tonnes of waste each year. Around 50 per cent is generated by construction, with 24 per cent from commercial and industrial activities and 14 per cent from households. The remainder comes from sources such as building operations, sewage sludge and farming.

Most of the waste produced ends up in landfill sites. It then generates methane, which contributes to climate change (see page 183). Also, a lot of energy is used up making new products to replace those thrown away. These new products later become waste themselves, contributing to further climate change. We are consuming natural resources at an unsustainable rate. We are also running out of space to put our waste. Suitable landfill sites are filling up and it is difficult to find replacements. There is now a tax on the disposal of waste. Active waste (waste that decomposes or is chemically reactive) is taxed more than inactive waste such as rocks, glass and concrete. In 2016, the tax was £84.40 for each tonne of active waste, a charge which is set to rise each year. Inactive waste is taxed at £2.65.

Figure 13 A landfill site in Belfast.

The waste hierarchy (see Figure 14) shows how reducing, reusing and recycling materials can help prevent waste.

Increased reuse and recycling are good for the environment – although not as good as reducing! Fewer resources and less energy are needed if reuse and recycling are done widely.

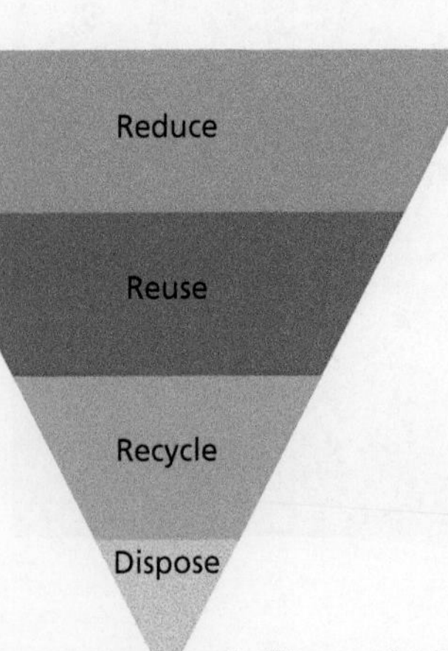

Figure 14 The waste hierarchy.

There are a number of reasons why dealing with waste is such a big issue in the UK:

- Shortage of landfill sites. Landfill is the term used for dumping waste into the ground, sometimes in old quarries or in natural hollows. This was how most waste was disposed of in the past. However, suitable sites are now much rarer as the old ones get filled up. In addition, people do not want to live beside landfill sites as they fear smells, blowing waste and vermin such as rats and seagulls.
- Environmental and health concerns. There are many chemicals in waste, such as mercury contained in batteries. If these chemicals reached the water in the ground they could poison it, killing wildlife and possibly people. Diseases too could spread in places where waste was rotting down.

The Government has set targets to reduce landfill and to increase recycling.

What is meant by 'reduce, reuse and recycle'?

You will often see the 'reduce, reuse, recycle' logo on items that should not be thrown away. However, not all of these actions with 'waste' are as good as the others. Reducing is best, and recycling is what you should do only if you cannot reduce or reuse.

REDUCE
REUSE
RECYCLE

Figure 15 The 'reduce, reuse, recycle' logo.

Reducing waste

Reducing is the best of the three options. This means reducing the amount of waste you produce, by buying less and consuming less. You can also reduce by choosing products that have less packaging than they need or that are packed in materials that have not consumed a lot of energy to produce. Other less obvious ways to reduce the amount of waste that you produce is to read a newspaper online, rather than buying one, and don't print out materials for your school work but download and read them online. Even turning off a tap while you brush your teeth is a way of reducing waste.

Figure 16 A sign encouraging people to reduce waste.

Reusing materials

If you cannot reduce, you can reuse 'waste' by finding innovative ways of using items again. You can use shopping bags again and again, instead of taking disposable plastic bags. These used to be free in shops but, since they are now charged for, their use has gone down and more people are using a 'bag for life'. In Northern Ireland, there were 190 million plastic bags used by supermarkets each year. After a 5 pence charge was introduced in 2013, plastic bag use fell to 30 million in just one year. In addition, the plastic bag charge in Northern Ireland raised £3.4 million in its first year, which was given to environmental groups. You can also reuse clothes by taking them to charity shops and other outlets, for others to wear. Furniture, toys, books and DVDs can also be passed along in the same way. Computer cartridges, leftover paint and computers can also be reused by others.

Figure 17 A bag for life.

Recycling materials

In the reduce, reuse, recycle hierarchy, the least useful way to dispose of waste is to recycle it. However, even though not as good as reducing or reusing, recycling is still better than letting the waste go to landfill. All councils in Northern Ireland collect waste for recycling from boxes or wheelie bins given to householders. All councils also have recycling centres for household waste – the householders that visit these centres separate waste so that they can be recycled easily. Most are equipped to recycle paper, glass and plastics, and some also collect metal and organic waste. Even small villages often have recycling banks and bins for glass or aluminium cans. In Northern Ireland, nearly two-thirds of what we call 'rubbish' can be recycled. It is said that recycling just one aluminium can saves enough energy (through avoiding the mining and smelting of aluminium ore, and transporting of new aluminium) to run a television for three hours.

While most people agree that the three 'Rs' of reduce, reuse, recycle are standards for sustainable living, there are additions suggested.

- Some suggest 'rethink', as a means to get people to think about how they could reduce the waste that they produce even further. They could ride a bicycle or use public transport rather than drive, for instance. Even when driving, they could avoid accelerating rapidly and check their tyre pressures regularly to avoid wastage of excess fuel. They might switch lights off when not required and ensure that taps are not left running, for instance.
- Another 'R' that has been suggested is 'refuse'. It looks a little like reduce, but refuse means that a consumer says no to something that they don't need or that might cause harm to the environment. You could refuse free samples that are given away by advertisers, or any items that you don't really need.

You could therefore end up with the hierarchy of 'rethink, refuse, reduce, reuse, recycle'. Only after that, would any item go to landfill. Landfill sites are costly. As well as the additional cost, putting waste into landfill is the worst option for the environment. Any resources within it are lost and the decomposing waste produces greenhouse gases.

Recycling of household waste across the UK has been increasing but the rate is slowing down and there are some doubts that the whole UK can reach its 50 per cent target for 2020. Even if it does reach 50 per cent, half of all the 'waste', much of which could be recycled or recovered, is still being thrown away needlessly.

Figure 18 Recycling bins.

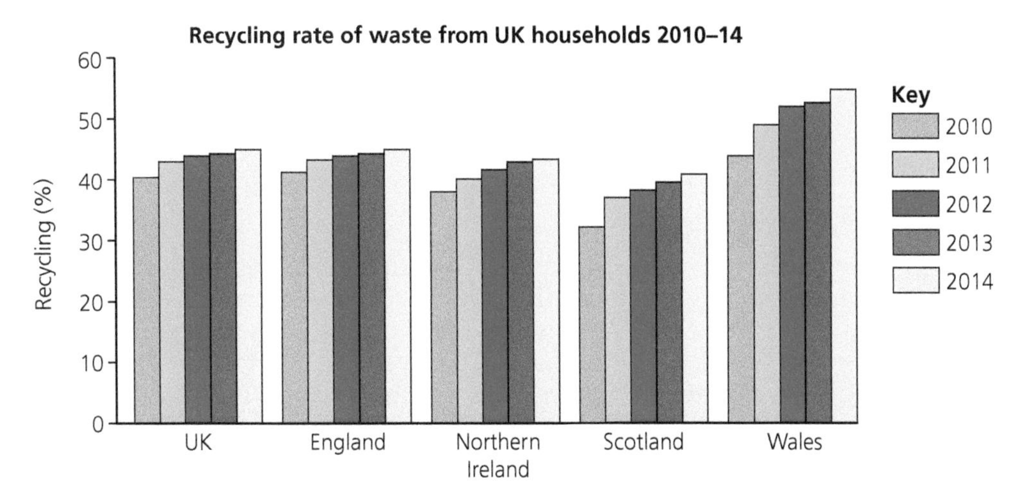

▲ **Figure 19** Recycling rate of waste from households, UK (Source: Office for National Statistics).

Activities

1 Working in a small group, plan a presentation for young people encouraging them to reduce, reuse and recycle. Your presentation must not involve any digital input. It might take the form of a song, or a role play. Show your presentation to other groups and view theirs.

2 There are a lot of myths about recycling which some people use as an excuse not to do it. Some of them are below.

a Prepare an answer to each of these myths, using figures where possible.
- Not recycling is cheaper than recycling.
- Since we have plenty of landfills, recycling isn't important.
- Someone goes through the rubbish bins and pulls out the recyclables before they go to the landfill.
- Only white paper is recyclable.
- It's OK to throw something away as long as it's biodegradable.
- Incineration is safe these days and you can burn rubbish into electricity.
- Recycling doesn't save trees or other natural resources.

b Work with others to produce a poster which challenges the myths.

3 Work as a class to develop a questionnaire about recycling to use with parents or guardians of those in your class.

a Decide what exactly you want to find out:
- Are you researching levels of recycling? Do these adults visit the local recycling amenity site? What do they recycle? Have they always recycled? Would they want to recycle more?
- Perhaps you are more interested in the attitudes to recycling. Do they think it is a good idea? Do they think it should be extended? How do they think the local recycling amenity site should be developed?
- Perhaps you are interested in their views about Energy from Waste plants. Do they know what they are? Would they support one being built in Northern Ireland? Would they mind if it was built close to where they live?

b If it is questions like this that you plan to research, and you might have different questions, then draw up a number of questions to ask. Make sure that you pilot the questionnaire before you use it. This would involve asking the questions of someone who is not going to be in the final survey to see if any of the questions are unclear or able to be misinterpreted.

By the end of this section you will be able to:

- evaluate the benefits and drawbacks of one renewable energy source.

A renewable energy source as a sustainable solution

What is renewable energy?

Renewable energy is energy which is collected from resources which are not going to run out on a human timescale. Oil, natural gas, peat and coal are fossil fuels which will eventually run out, so these are not renewable. Harnessing the power of the wind with wind turbines creates wind energy, a renewable energy. Biofuels are made by converting plants such as maize, wheat or sugar cane into energy. Capturing energy from deep in the Earth is another renewable source of energy. Further renewable sources are capturing wave energy or the energy from tides, or generating energy through Hydroelectric Power (HEP).

Solar energy

A further promising renewable resource is the use of the Sun's energy to create electricity that people can use – solar energy. There are two main ways to capture this energy. The first is by Concentrated Solar Power (CSP), which focuses the power of the Sun.

▼ **Figure 20** A parabolic trough.

▲ **Figure 21** A Power Tower.

Parabolic troughs (see Figure 20) are curved mirrors that focus the Sun onto a tube held above the trough, on which the Sun's rays are concentrated. In the tube is a fluid such as synthetic oil which absorbs the energy, reaching temperatures of over 400°C. This heat is used to turn water into steam and the steam drives a turbine to create electricity.

Another Concentrated Solar Power system uses flat mirrors which are controlled by computers to track the Sun and focus its rays on a central tower. In this Power Tower system (See Figure 21), the liquid is heated to even higher temperatures than the parabolic troughs, but it produces electricity in the same way.

The second main way to generate electricity from the Sun is to use photovoltaic cells. The solar panels seen on roofs in the UK are mainly of that type. Large-scale generation is achieved from large arrays of photovoltaic cells in solar farms. These turn the Sun's energy directly into electricity and can be permanently mounted or built to track the Sun as it moves through the sky.

Photovoltaic cells used to be very expensive to produce but costs have fallen because of advances in technology and increases in the scale of production.

Figure 22 A photovoltaic array.

Much of North America has deserts which are ideal places to generate solar power because of the amount of sunlight which they receive. Australia had been criticised for using little of its potential for solar power, but that seems to be changing rapidly, and it is now among the world's ten top solar countries. Europe is increasing the amount of solar power it generates too. Concentrated solar power is generated mainly in Spain, while photovoltaic cells are used in most countries in Europe, with electricity production increasing all the time. Solar electricity generation is also increasing in Africa, Central and South America and Asia.

What are the benefits of solar power?

There are many advantages of solar power.

- These are non-polluting technologies when in operation, producing no noise and no greenhouse gases. They have to be manufactured and transported to the sites where they will be used, so there is some pollution, but much less than using fossil fuels, for example.
- Once established, solar plants are cheap to run, requiring only a little maintenance.
- Photovoltaic cells have no moving parts so will last a long time. Typically, they are guaranteed for 25 to 30 years, but most have a life expectancy of 40 years.
- This allows electricity to be created locally, saving on pylons and wiring. This assumes a sunny location where the electricity is required.
- It creates employment. The solar industry claims that 33,500 people were employed in solar energy across Europe in 2010.
- The use of solar energy avoids the price fluctuations that oil and natural gas experience.
- It increases fuel security for those countries with adequate sunlight.

What are the disadvantages of solar power?

There are some disadvantages, however:

- Electricity cannot be created when there is no sunlight, although some technologies store the energy, for instance in super-heated molten salt, which is then used to generate electricity even at night.
- Production is reduced during cloudy periods, storms and in winter, and it is difficult to store the electricity for these periods.
- Even the most efficient photovoltaic cells convert only 20 per cent of the Sun's energy to electricity, meaning that a large area is needed to generate the power required, although the efficiency is increasing as technology improves.
- Solar panels are bulky for domestic use, although new technologies are reducing these.
- The NIMBY (not in my back yard) effect means that some applications for solar farms, including in Northern Ireland, raise objections from those who see them as an eyesore.

Sometimes electricity can be generated in one country and transported to another, and this can bring other advantages and disadvantages. Many of the countries of the Middle East and North Africa (sometimes called MENA) have tremendous potential for generating renewable energy, particularly solar power, in their deserts. This includes Algeria, Morocco, Tunisia, Egypt and Jordan. They also have high unemployment, and often low GDP, and some are dependent on the import of fossil fuels, particularly oil, to provide power for their growing populations. The potential for solar energy in North Africa alone is said to be the same as producing 1 million barrels of oil each year.

Since 2009, there has been a plan to tap the potential for solar power (see Figure 23) from MENA countries, allowing them to export some of their power to Europe along interconnectors. These would allow, for example, solar energy to be exported from Algeria to France, and from Libya to Turkey and Greece. This cheap and green electricity could then be transported to the cities and industries in northern Europe, including in the UK and Ireland. This project is ambitious and, at an estimated €400 billion, costly.

Figure 23 A map showing the potential to create solar power.

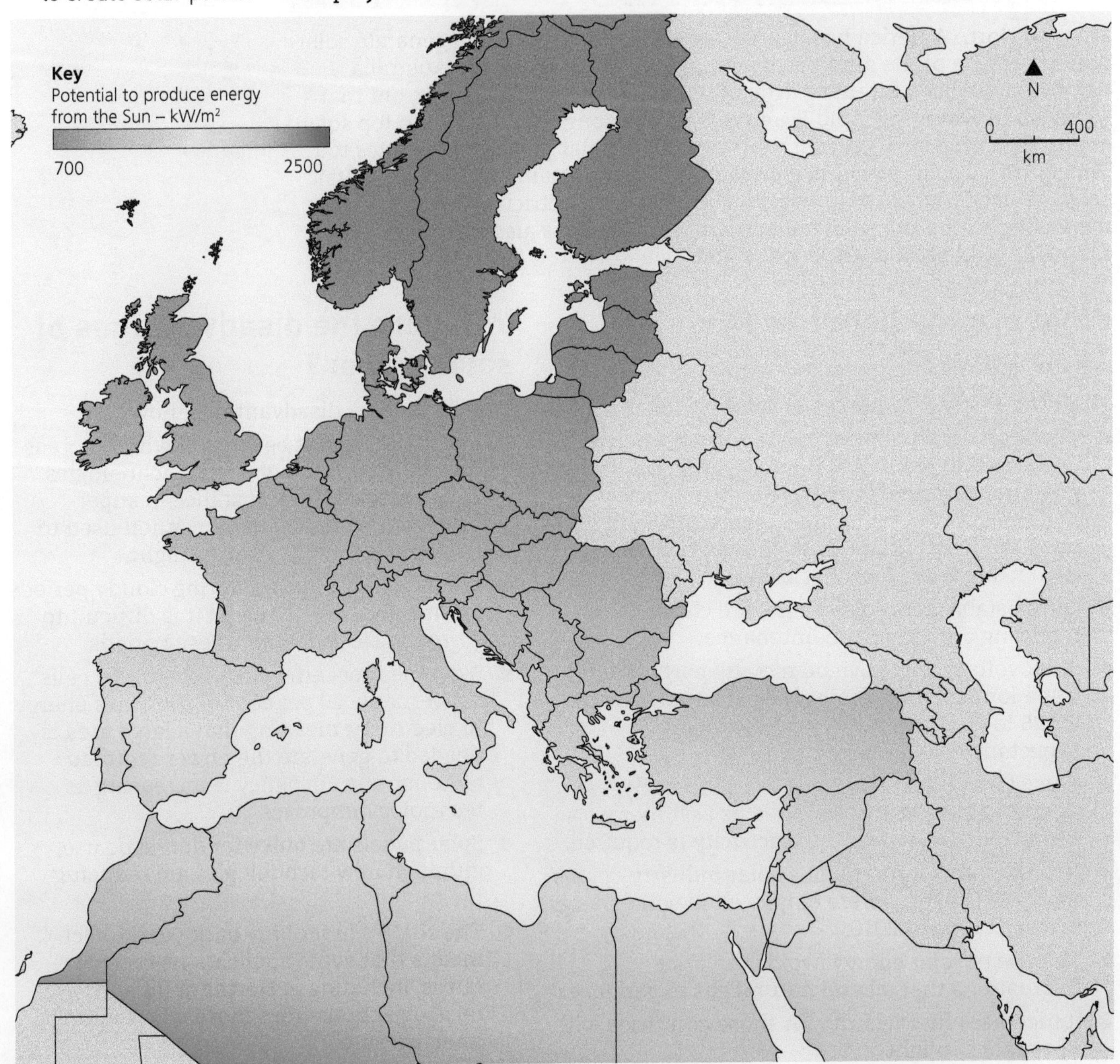

There are several advantages and disadvantages of developing solar power in MENA countries.

Advantages	Disadvantages
▶ Providing a locally generated renewable power source to drive local industry in the MENA countries ▶ Providing short-term employment, particularly for young people, in constructing the plants ▶ Providing long-term employment in operating the solar plants, and in the industries which could grow up in the MENA countries ▶ Increased earnings from exporting power to Europe ▶ This clean energy will be cheaper to produce than that being produced in Europe currently ▶ The power will be available in every season of the year – there is no 'down time' ▶ The development will contribute to taking carbon out of the atmosphere by replacing fossil fuels with renewable energy ▶ Europe could gain ten to fifteen years in the fight against climate change	▶ These are enormous large-scale projects which will bring few benefits to the local communities ▶ Any benefits are largely for the MEDC countries – other smaller-scale schemes would benefit MENA countries more ▶ The technology is MEDC technology, and will be owned and controlled by them ▶ There is a mistrust of multinational corporations, who are known for their exploitation of the countries in which they work ▶ This has more to do with freeing Europe from dependence on Russia's energy exports than on helping MENA countries ▶ The plants need water for cooling and to clean the mirrors, and these countries have a shortage of water, made worse by climate change

The plan seems to be on hold at present, but the MENA countries are building solar farms in any case. One of the largest is Ouarzazate in Morocco, the first phase of which started producing electricity in 2016. It will, by 2018, produce enough power for 1 million homes and reduce carbon emissions by around 760,000 tonnes each year. There has been an investment of US$3 billion, which includes US$1 billion from a German bank. Morocco's electricity demand was growing at 7 per cent a year and the country had to import 97 per cent of its energy. Morocco has pledged to generate 42 per cent of its power from renewable sources by 2020, and this helps in that goal. However, some Moroccans are concerned about the increasing control by multinationals on electricity production in their country.

➤ **Figure 24** Ouarzazate solar farm.

Activities

1 Open GoogleEarth and enter the co-ordinates +31.008801–6.861589 into the Search field and press return. This will take you to the CSP plant in Ouarzazate.
  a Use the measuring tool in GoogleEarth to calculate the size of the plant shown in the image.
  b Use the History tool to look at images taken in the past. Can you work out when it was started?
  c Try to identify the central part of the electricity plant where the steam is used to generate electricity. Can you spot the same place in Figure 24? What direction was the camera facing to get the image in the book?
  d Is there evidence of Phases 2 and 3 starting? What is that and where? Use the measuring tool to calculate what size the solar farm will be, eventually.

2 Print out a screenshot of the solar plant and put it in the centre of a large piece of paper, or a digital presentation.
  a Around it put the advantages and disadvantages that solar power might have, in general.
  b In a different colour, put in some of the advantages and disadvantages that this particular plant might have.

By the end of this section you will be able to:

- evaluate the 2015 International Climate Change Agreement (ICCA).

The 2015 International Climate Change Agreement (ICCA)

The 2°C degree rise in global temperatures (see page 182) is generally agreed as the 'tipping point' among scientists involved in climate change research. After that point, it will be impossible to prevent serious implications for the whole planet.

What is the 2015 International Climate Change Agreement (ICCA)?

This is an international deal on climate change held in Paris, which was signed up to by 195 countries around the world and finally agreed in November 2016. It is also called the Paris Agreement. This is the first arrangement which commits all countries to addressing climate change, and the United Nations spent nine years negotiating to get this agreement. Barack Obama, the United States President at the time, said 'This agreement sends a powerful signal that the world is fully committed to a low-carbon future', and the then United Nations Secretary General, Ban Ki-moon, warned that there was no alternative if the deal fell apart. There is 'no Plan B because there is no planet B', he said. For the first time, the world seems to be united on this issue.

Reducing emissions

The conference agreed a long-term goal of keeping global average temperatures well below 2°C above pre-industrial temperatures. They also recognised the importance of reducing the risks and impact of climate change by limiting the increase to just 1.5°C. They agreed that the growing global emissions of climate changing gases such as carbon dioxide should peak as soon as possible, while recognising that this will take longer for LEDCs. There was agreement that the emissions should reduce as quickly as possible after that peak. Each country came up with their own action plans for reducing their emissions.

Figure 25 Welcoming the 2015 International Climate Change Agreement.

How is it to be monitored?

The countries agreed to get together every five years to set ever more ambitious targets. They also have to report to each other and to the public on their progress with their action plans in a global stocktake. The whole process is to be transparent so that all countries can see the progress being made by other countries.

How will countries adapt?

The governments agreed to increase the ability of societies to adapt to the negative impacts of climate change. LEDCs in particular would get international support to adapt. MEDCs would continue with a goal of providing US$100 billion per year by 2020, extending this to 2025, with a new and higher goal to be set after that point.

The conference also recognised the importance of preparing for climate change through averting and minimising loss and damage from it. There was a recognition that this would involve co-operation in areas such as early warning systems, emergency services and risk insurance.

This all sounds complicated but, for the first time ever, it demonstrated an acceptance by the world's countries that ecosystems, biodiversity and land use can be used to reduce greenhouse gases and help countries and communities within them to reduce risks and adapt to climate change. This might involve, for example, conserving and restoring forests. It might equally involve changing some agricultural practices. Some people have described this as 'calling out the power of nature'. It shows a realisation that healthy ecosystems protect people from the effects of climate change.

There are two main processes at work in the agreement.

- Stocktaking: countries will announce publicly every five years how much they have reduced carbon emissions. This transparency is a key feature of the agreement. There is still discussion as to when the stocktaking will occur, with most countries arguing for 2023 or 2024, and taking place every five years after that.
- Ratchetting: each target set will be replaced by a more ambitious one at regular intervals. The time between these is still being discussed. India wants it to be every ten years, as its short-term goal is to raise the standard of living for its population. Some other countries such as the USA and Pacific Island nations want the targets to be reset every five years.

Are there any problems with the deal?

Some people argue that there should have been an outside agency, perhaps like the International Atomic Energy Agency, to make sure that countries kept their promises to cut emissions. India, China and other LEDCs were sceptical of that arrangement so instead all countries will have the same framework of transparency, although the agreement recognises that LEDCs may need more help in reaching their targets.

Aircraft emissions are the fastest growing source of pollution, and these are emitted at high altitudes where they cause most damage to the atmosphere. These were not part of the Paris Agreement, but were agreed separately. Aircraft companies were allowed to increase emissions from international flights until 2020, but after that have to limit their carbon production to those levels. If they exceed the 2020 cap, they have to compensate by, for example, planting trees. Trees store carbon, so allowing more carbon to be emitted from aircraft. However, many environmentalists and others fear that the agreement on pollution from aircraft is too easily avoided by the aircraft companies. Who will monitor the tree planting, and make sure that areas planted are only counted once, for instance? What if the trees are eventually felled and their carbon released? One commentator called it a 'game which does little more than help airlines hide their rapidly growing threat to our climate'. Others have pointed out that it will eliminate about 2.5 billion tonnes of emissions.

Are countries told what they have to cut and when?

No. The countries that have signed up to this agreement can independently decide on how they will lower their emissions. This is very different from earlier attempts to get climate change agreements. Earlier attempts required all signing parties to adopt similar measures. However, in this agreement, there is an acceptance that economies, cultures and nations differ so greatly that it would be very difficult to get measures that everyone could adopt. This allowed all the 195 countries to decide the best way forward for them, individually, and this meant that it was easier for all countries to support the agreement.

In other words, the Paris Agreement has a 'bottom up' structure. This is different to most international environmental law treaties which are 'top down', laying down targets and standards and for states to implement.

Critics of the deal

Some are concerned that the agreement is based on reducing emissions voluntarily. Professor James Hansen, a climate change expert, is angry that most of the agreement consists of promises or aims and not firm commitments. Others too are concerned that there is no binding enforcement mechanism. The biggest polluters which generate more than half the world's greenhouse gas emissions (such as China, the USA, India, Brazil, Canada, Russia, Indonesia and Australia) are being expected to measure and control their own emissions without any penalty if they do not do it properly, or if they do not reduce their emissions.

In some parts of the world there are still some who, despite the overwhelming scientific evidence, are sceptical that climate change is caused by people, and some of them have argued that the climate change agreement will hurt the economies of their countries. For others, even if it caused a reduction in economic growth, it is a price worth paying to protect against the effects of climate change across the world.

Figure 26 The Paris Agreement is seen by many as the best hope there is to preserve the planet.

Activities

1 As a geographer who has studied climate change, you are tasked with persuading some politicians who are sceptical about climate change to accept that it is a serious threat. Work in groups to prepare a short statement to the politicians. You should try to be as persuasive as possible, and use facts and figures to support your arguments. Think about what might make the politicians change their minds, and try to address that.
Present your arguments. Which group had the most convincing presentation? Why was it the most convincing? What would you change if you wanted to be more convincing the next time?

2 Prepare an infographic to highlight the 2015 International Climate Change Agreement.
   a Before you begin agree on success criteria for the infographic. Should it have:
      - a bold and catchy title?
      - text to inform the reader?
      - images – what kind?
      - figures?
   b When the posters are completed, they can be judged against these criteria.

What is mass tourism and how has it grown?

By the end of this section you will be able to:

- evaluate the positive and negative impacts of mass tourism
- understand how to be a responsible tourist.

Tourism has grown very rapidly in recent years. Before that growth, most tourism was local and on a small scale. In Northern Ireland, for example, the tourist resorts of Bangor and Portrush grew up in Victorian times and were linked to the main cities by train. At that time most working people spent their one-week holiday in a seaside resort. Other holidays away from home were rare, and travelling abroad was even rarer, except by the very wealthy. This started to change in the 1960s.

A number of developments started then that transformed tourism into mass tourism, which has had an impact on the whole world. The changes included:

- Increased leisure time. People began to get longer holidays. For example, since the mid-1960s, the typical UK adult is now working almost eight hours less per week, which is the equivalent of between seven and nine weeks' holiday each year.
- Cheaper travel. Travel is increasingly more affordable for more people. Cheap air travel has brought what were once faraway destinations into reach for many people. The internet has helped to bring down the cost of travel and holidays and more people are booking their holidays online.
- Increased disposable income. Alongside shorter working weeks, in 2005 UK workers earned four times as much as they did in the 1980s. People therefore have a larger proportion of their salary available for their holidays.
- Increased health and wealth of older people. As people are living longer and healthier lives, in general, they are now able to travel more than they could before. This has extended the season for many traditional summer resorts in the sun as older people take long holidays in the winter, avoiding the worst of the cold weather at home.

Tourism in the UK and Ireland is now characterised by mass tourism. This is mainly to Spain, Greece and Portugal. However, there is some evidence of this pattern changing, with tourists rejecting package holidays and increasingly travelling further. There is now nowhere on the globe which is out of reach of tourism and its impact.

Figure 27 A crowded beach in the Algarve, Portugal.

What is the impact of mass tourism?

Tourism is an enormous industry, growing at 6 per cent every year. In terms of the numbers employed, it was estimated to be the biggest industry in the world in 2000. Over 200 million people work in the industry, and earn 11 per cent of global earnings.

The World Tourism Organisation calculated that there were 903 million tourists and that tourism earned £582 billion in 2007. In 1996 the earnings were just £258 billion. The number of international tourists is expected to double by 2020, reaching around 1.5 billion.

Tourism exploits the attractions of an area. Some of the attractions might be scenery, isolation, unspoilt natural beauty, wildlife or welcoming local people. The dilemma is that, when the tourism resources of an area have been consumed by large numbers of tourists, the original features which attracted people to the area have often been destroyed. Tourists will then move on to the next 'undeveloped' location.

Increasingly, it is understood that this is unsustainable, and efforts are being made to move towards a form of tourism which does not destroy the places they visit – this is sustainable tourism.

Cultural impact

Tourism can have a beneficial impact on the culture of an area, opening it up to other cultures and helping people to understand each other better. It can generate in local people a pride in their history and culture and in their buildings and environment. In some places tourism, preservation, heritage and culture are much more likely to overlap.

Tourism can also damage local cultures. Tourists may challenge the traditional attitudes to work, money and relationships. The restaurants, bars, discos and other entertainments provided to attract and entertain tourists can bring disturbing public behaviour, including drunkenness, vandalism and crime. Young locals may copy the visitors' behaviour and this can result in social conflict.

▲ **Figure 28** Graffiti at Casa di Giulietta, Verona.

Figure 29 Hawaiian 'culture' on display for tourists.

Some tourists travel to places to take 'selfies' which involve them taking photographs of themselves in exotic locations. There has even been a trend for young MEDC tourists to pose for naked selfies when visiting temples and other sacred sites, which local people find very disrespectful. Tourists have defaced many famous sites, including ancient Egyptian temples, the House of Juliet in Verona (see Figure 28) and some have even scratched messages on the bunks in Auschwitz.

Some locals are encouraged to dress in their traditional costumes and perform rituals for tourists. This can devalue local customs and reduce real people with a very rich culture into something for tourists to photograph. In some cases what is left of the culture is so geared for tourism that most of the meaning of the locals' lifestyles has been lost. One example is in Hawaii with, for example, the traditional greeting by 'hula girls' at Honolulu airport. The spiritual and sacred nature of traditional Hawaiian dancing has been lost. One writer quotes a native Hawaiian as saying: 'We don't want tourism. We don't want you. We don't want to be degraded as servants and dancers. I don't want to see a single one of you in Hawaii.'

Economic impact

We have seen that there are enormous amounts of money earned through tourism. It has a considerable economic impact in the areas where it is operating, and employment in tourism can generate a considerable income for local people, particularly if there are few alternative means of employment. This money can act as a stimulus to develop other forms of work and bring even more money into the area.

However, much of the money earned through tourism does not stay in the place where it is spent. If tourism is controlled by large companies, not based locally, much of the money spent will go outside the area. For example, a visitor to Belfast might travel with a tour company which is run by a Spanish company, stay in a hotel owned by an American chain, eat in restaurants owned by an English firm and buy gifts which might be made anywhere in the world. If this were the case, very little of the cost of that holiday is left in Northern Ireland to stimulate economic growth. There is employment, of course, but many of the often poorly paid workers involved in tourism, such as waiters, porters and maids in hotels, may be migrant workers, so even there much of the potential benefit may be lost. Also, employment in tourism is often seasonal so jobs may only be available for a short tourist season.

Environmental impact

Tourism can improve the quality of the environment in an area. The money earned can be spent on improving roads and other transport links, and on building better houses and public buildings, for example. It can also be invested in improving treatment plants for water supplies and for sewage. The features that tourists are coming to see – whether scenery or wildlife or historic buildings – are more likely to be conserved as a result of tourism.

However, tourism can also have negative effects on the environment. Water is a particular issue. Water is an important resource for tourists who require such things as swimming pools, golf courses and landscaped gardens. Water is also needed to support those industries which provide services for tourists, such as agriculture. Water also has to be of a reasonably high quality as tourists expect certain standards of purity for drinking, bathing and for recreational use. A survey in Zanzibar, a popular tourist destination on the east coast of Africa, showed that tourists, on average, used 685 litres of water each day; local residents use less than 50. Tourists staying in hotels rather than guesthouses used even more – an average of 931 litres each per day. Sustainable water management would suggest a limit to consumption of between 200 and 300 litres.

This has an impact on the environment as this water has to be removed from the ground or stored in reservoirs and treated before use. Often the water is also needed for agriculture and this may leave local people short of this precious resource.

Often tourism increases the pressure on local ecosystems. Massive numbers of tourists have a huge impact and put local habitats at risk. The tourists produce waste and pollution which can affect the local wildlife. Some tourist resorts feed sewage and other waste directly into the sea, damaging surrounding coral reefs and other sensitive marine habitats. To create open beach sites, important ecosystems such as mangrove forests and seagrass meadows are removed. Large numbers of tourists on beaches and lit-up seafronts have endangered marine turtles, which rely on dark secluded beaches to lay their eggs.

The increased popularity of cruise ships has also affected the oceans. With up to 4,000 passengers and crew on board, these massive 'floating towns' are a major source of marine pollution. The waste produced on these cruisers is dumped overboard and sewage is discharged into the ocean, along with all of the other pollutants which large ships produce.

How can you be a responsible tourist?

Thirty years ago it was recognised that tourism can cause damage and destroy the very things that people are coming to see, including the environment and the scenery. The result was an increasing focus on ecotourism. Tourists were told 'Take only photographs, leave only footprints'.

It is now recognised that ecotourism is not enough, because often that does not recognise that there are people living in tourist destinations, and tourists should try not to damage not only the environment, but also the cultures and societies of the people who live in the places they visit. This has been called responsible tourism.

A responsible tourist should be concerned about protecting and preserving places, cultures and environments, to the benefit of both the traveller and the host. The Ethical Travel Portal publishes a

Figure 30 Cleaning a beach in the Philippines on Earth Day.

list of things that you should do if you are going to be a responsible tourist. These are some of them:

Before you travel:

- Do some research into the place you are visiting and learn a little of the local language and the customs.

And when you are there:

Social responsibility

- Remember that you are a guest and be respectful of the local customs
- Ask before taking photographs
- Be polite

Economic responsibility

- Try local food in local restaurants
- Buy local food and contribute to the local economy
- Use local transport where possible
- Buy locally produced goods and services

Environmental responsibility

- Be careful with your use of resources such as water
- Don't use towels more than necessary
- Don't pick wildflowers
- Don't leave litter behind
- Don't buy products made from endangered plant or animal species.

Some of the items on this list are about ecotourism, but the others make this responsible tourism.

Being a responsible traveller does not mean that you have to holiday in a eco-lodge high in the rainforest canopy. Any tourist can be a responsible tourist, whether they are going to a common destination or an exotic one. The key thing is to respect the environment of the place that you visit and the people who live there. Responsible tourism can involve staying in locally run hotels and guesthouses, using independent shops and restaurants and visiting nature reserves where your entrance fee pays towards conservation. It aims to make tourism a more enjoyable experience both for the tourist and for the local person.

As with all claims to be 'environmentally friendly', or 'eco friendly', tourists have to be very careful that the claims that are being made for responsible tourism are not just 'greenwashing'. This term is modelled on 'whitewashing' which is the covering up of something that you don't want to be seen. Greenwashing is when companies try to mislead the public by pretending that their products from washing powder to vehicles, but also including holidays, are more environmentally friendly than they actually are. Putting on a green spin in advertising, starting the brand name with Eco, or claiming products as 'natural' are strategies often used. Responsible tourists would want to be sure that the claims were not 'spin' before they would go on a holiday which was said to be 'environmentally friendly' or 'responsible". Responsible tourists would want to be sure that the claims were not 'spin' before they would go on a holiday which was said to be 'responsible'.

Activities 

1 Below are ten activities that tourists do.

a Which of the actions are those of a responsible tourist, and which are not? Is it always easy to decide?

- Pay a local guide to show you a local cave system
- Visit churches while wearing swimwear
- Stay in the Hilton Hotel and eat in McDonalds
- Buy expensive imported food and drink
- Learn to say hello, goodbye and thank you in the local language
- Take photographs discretely of local people at their work
- Arrange your travel with a reputable multinational travel company
- Use water carefully, wasting very little
- Stay in an environmentally friendly hotel, owned by a company from a different country
- Eat the foods that local people eat

b Think of another few activities that would show what is 'responsible' and what is not.

2 Mass tourism impacts on the environment, economy and the culture of the places where it happens. Some of these are benefits, while others are negative. Working in a group, research the impact of mass tourism.

a Arrange what you find out in a poster around the heading MASS TOURISM. Using colours, code the positive and the negative effects. For each negative effect, suggest a way responsible tourism could address the problem.

b If all tourists were responsible tourists, would many of these problems be reduced?

c Agree a summary sentence for your poster and add that at the base.

By the end of this section you will be able to:

- describe and explain ecotourism
- assess how ecotourism can protect the environment.

Ecotourism

Ecotourism is a type of tourism involving visiting fragile or relatively undisturbed natural areas. It is intended to be a small-scale and low-impact alternative to mass tourism.

What are the principles of ecotourism?

Ecotourism tries to combine:

- conservation
- communities, and
- sustainable travel.

If it is to work, those involved in ecotourism, whether the firms who put together and market the holidays or those that take part in them, should:

- reduce as much as possible any impact, whether on the environment or the people
- design, construct and operate tourist facilities which are low-impact
- respect the environment and culture of the places visited
- provide positive memorable experiences for visitors which help to raise sensitivity about host countries
- generate income which is aimed at conservation
- generate income both for local people and for private industry
- recognise the rights and beliefs of the indigenous people in the community
- work in partnership with local people to give them a role in the tourism.

Figure 31 A sign in Guatemala saying, 'Go easy! This is an ecotourist area'.

How can ecotourism protect the environment?

CASE STUDY

Ecotourism in Nam Ha, Laos

- Nam Ha is an area in the province of Luang Namtha in Laos, a country in South East Asia. The official name for the country is Lao People's Democratic Republic. Nam Ha covers 2,224 km^2 from the lowlands of the Luang Namtha Plain to mountains of over 2,000 m in altitude. The area was protected by being declared a National Biodiversity Conservation Area in 1993. Twelve years later it was declared a Heritage Park by ASEAN (Association of Southeastern Asian Nations). It won the Equator Prize in 2006. This prize is awarded by the United Nations Development Programme every two years in recognition of outstanding community efforts to reduce poverty through the conservation and sustainable use of biodiversity.

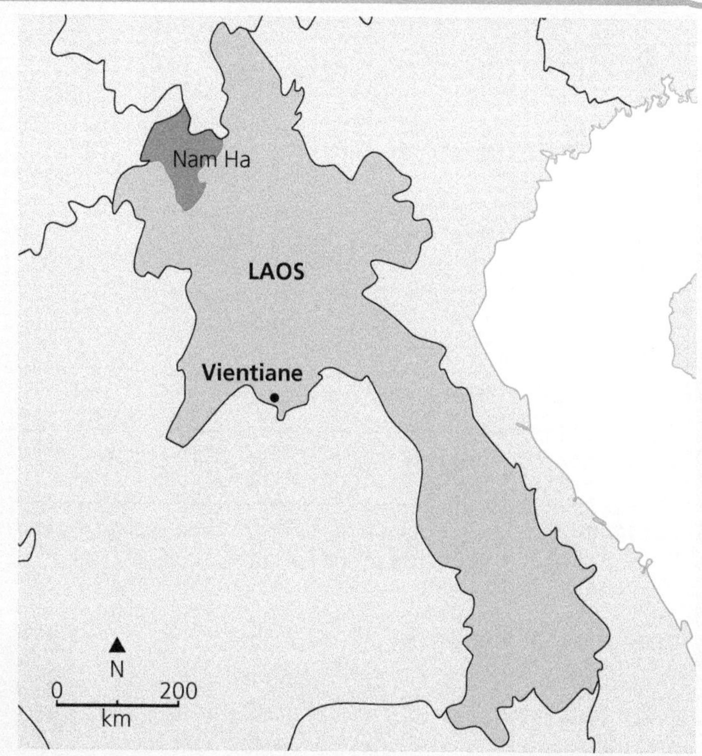

Figure 32 Map of Nam Ha, Laos.

The Heritage Park

- The park includes some of the largest and most important wilderness areas in Laos. Over 90 per cent of the protected area is covered by dense mixed deciduous forest. Because of the variety of habitats, it supports a rich variety of plants and animals living in the wide range of ecosystems. This biodiversity is of international importance.
- There are at least 37 larger mammal species found in the Heritage Park. Of these a number have been identified as especially important for conservation. They include large cat species such as the clouded leopard, leopard and tiger, all of which are very rare. Leopards are hunted for their fur and suffer from loss of natural habit due to the spread of people in many parts of the world. The WWF has warned that the world's tigers are in danger of extinction, with numbers having fallen by half over the past 25 years. There are thought to be as few as 3,500 left across the planet. There are many other rare animals too, such as gaur (Indian bison) the endangered Giant Muntjac deer, and a small number of Asian elephants. Birdlife is equally rich with at least 288 species recorded in the area. Of these, a number are very rare and need conservation, including types of pheasant, woodpecker and the endangered Giant kingfisher. The smaller animals in the area have not yet been studied but it is likely that these too are rich and diverse.
- An ecotourism project organised by UNESCO is currently working in Nam Ha. The Lao Government's tourism authority is working closely with the Ministry of Agriculture and Forestry, the Department of Forestry Resource Conservation and the Ministry of Information and Culture. UNESCO provides some help in operating the project, which is funded by grants from New Zealand, Japan and other sources. The area was identified by the Laos Government as having a high potential for both cultural and nature tourism.
- The people who live in the area have rich cultures and are of interest to visitors. The project provides economic opportunities for local people as the ecotourism project includes 57 villages, home to 3,451 families comprising a total of 21,227 people. The locals are trained as eco-guides and they offer village-based lodges and forest camps to tourists. They are also trained to monitor biodiversity, supporting the government officials whose responsibility that is, but who are under-staffed and poorly resourced.

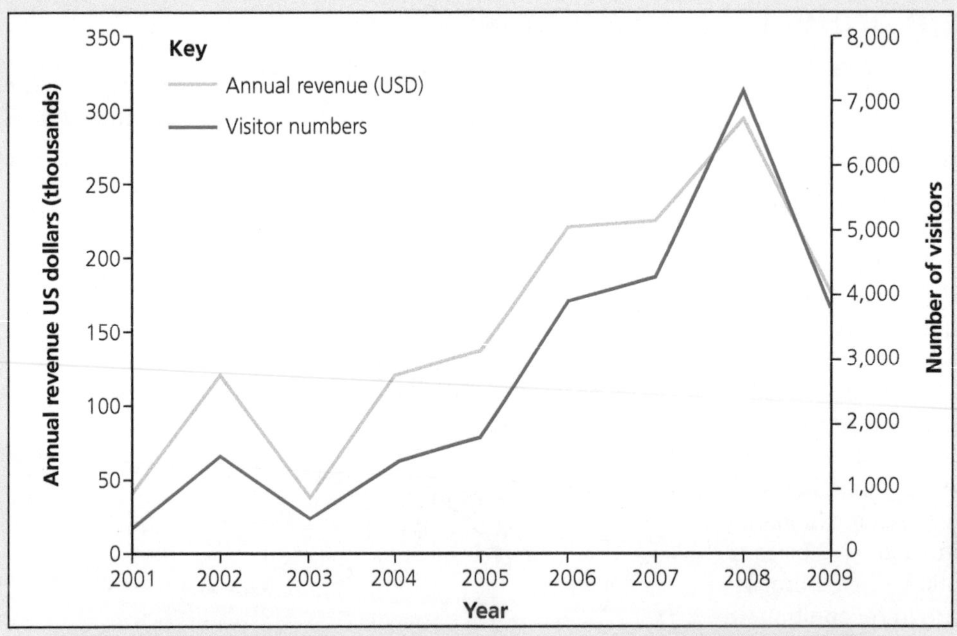

▲ **Figure 33** Nam Ha Ecotourism Project visitor numbers and revenue 2001–09 (Source: Nam Ha Ecotourism Project).

Activities in the park

There are a range of activities ranging from boating on the Nam Ha and other rivers in the Park to trekking around the area. There are also caves in some areas which will attract tourists.

Aims of the Nam Ha Park

The project aims to make sure that the tourist industry in the Nam Ha area:

- contributes to the conservation of the natural and cultural heritage of Laos
- involves local communities in the development and management of tourism
- uses tourism as a tool to promote development in rural areas
- provides training and skills to local communities to help fight poverty
- produces tourism which is sustainable, both culturally and environmentally
- helps communities to establish cultural and nature tourism activities in and around Nam Ha.

Challenges

The dilemma for the Laos Government is how to promote tourism, with all the benefits that would bring for the inhabitants of Nam Ha and the whole of Laos, without damaging the wildlife and beauty of the area.

▲ **Figure 34** Nam Ha Heritage park.

Activities

1. In groups think about the following question: In an LEDC, like Laos, how might the government promote ecotourism in a region like Luang Namtha province, balancing the damage that it might do to the region's ecosystems with the benefits that local people could receive?
2. Use a search engine to find out information about the ecotourism there. You might use search terms such as:
   - ecotourism in Laos
   - Nam Ha
   - Laos tourism
   - Luang Namtha tourism.

   Remember to check the source of the information that you find, and be cautious if you think it is not from a source which would be reliable, or which might be unfair in what it says.
3. Prepare a digital presentation to give the views of your group to the rest of the class. It is important to include visual information such as maps and photographs, as well as graphs and sketches and diagrams where they help to 'tell the story'.
4. Give your presentation to the class. Everyone in the group should be involved in delivering parts of the presentation. One key rule: don't read off the words on the screen!

Research activities

There are also tourism opportunities in MEDCs which try to be ecofriendly. One example is Hillwalk Tours (www.hillwalktours.com), which organises hiking tours all over Ireland and the UK. The tourists are provided with local accommodation, with the income benefiting local communities.

Working in groups, your task it to find and research an example of ecotourism. Be alert for `greenwashing'. Then present it back to the rest of the class in a Prezi (www.prezi.com).

As a group you need to decide:

- the main focus for your presentation
- the content areas that need to be researched
- the type of visual material needed
- how to present your research findings
- the roles for your group members.

Sample examination questions

1 What are biofuels? [2]

2 Study Figure 1, which shows the waste hierarchy, and answer the question below.

Why is 'dispose' much smaller in the diagram than 'reduce'? [2]

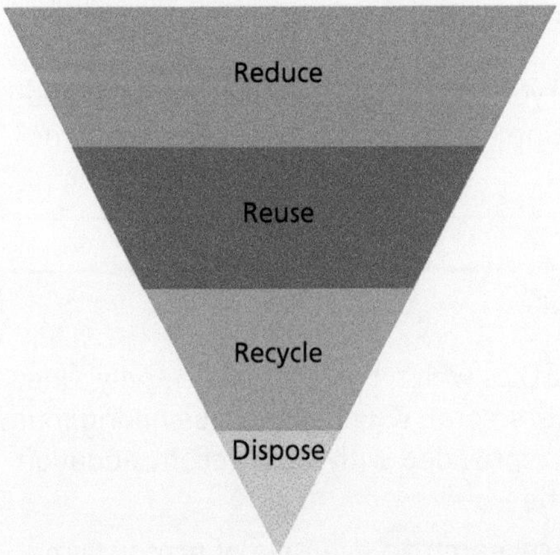

Reduce
Reuse
Recycle
Dispose

Figure 1 The waste hierarchy.

3 Climate change is increasingly recognised as having an impact on people, the economy and the environment.

Describe one benefit and one disadvantage of climate change on the environment. [4]

Benefit:

Disadvantage:

4 A website for responsible tourism (www.responsibletravel.com) says 'our idea for responsible tourism was that it should ... deliver "better places to live in and to visit"'.

Describe and explain the social, economic and environmental actions that a tourist should take to be a responsible tourist. [6]

Figure 2 An ecotourist lodge in Costa Rica.

5 Ecotourism is a growing form of tourism around the world. Describe and explain 'ecotourism'. [5]

6 Outline the main recommendations of the 2015 International Climate Change Agreement, and evaluate how effective it might be. [6]

UNIT 3

FIELDWORK

A fast-flowing river.

Is this a safe river to investigate?

1 The geographical enquiry process

Fieldwork is a vital part of studying geography. It allows you to test ideas in the real world and can help you to gain a deeper understanding of a geographical topic. It also aids the development of flexible skills that employers look for, such as teamwork, data analysis, interpreting data from graphs and evaluating evidence.

In GCSE Geography you will need to collect data related to a topic you are studying. To be able to sit Unit 3 you will have to produce a fieldwork statement and table of data. The fieldwork statement should include:

- title
- aim
- hypotheses
- details of the location of your study (you may include a map).

The process of collecting the data, writing up results and analysing data will help you become a more confident geographer.

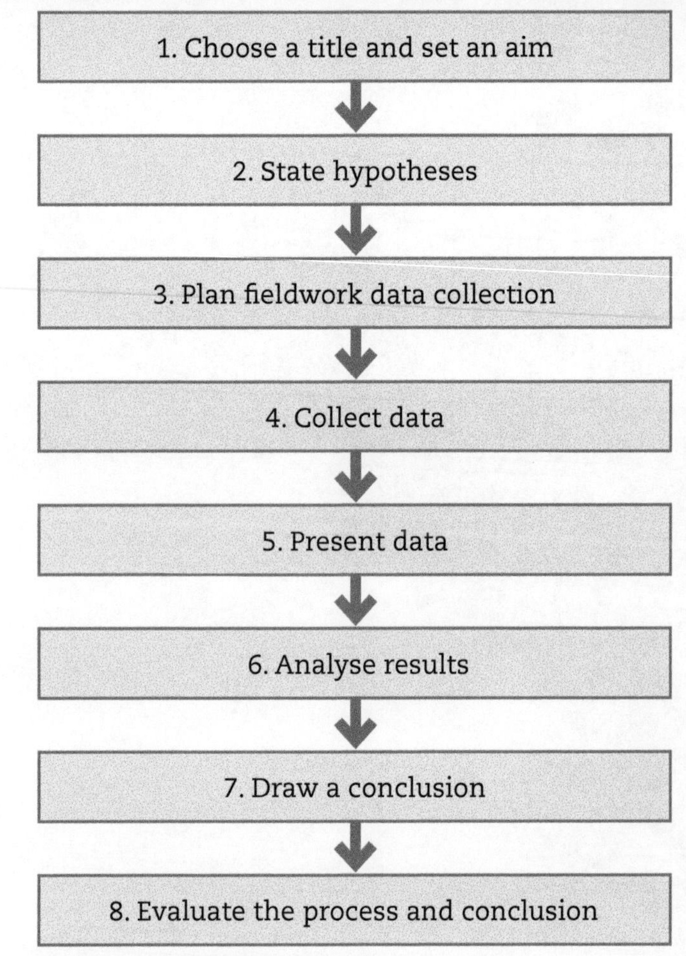

Figure 1 The steps needed to carry out a geographical enquiry.

Figure 2 A student using a caliper during a river fieldwork study.

How are geographical enquiries carried out?

Most geographical enquiries follow a set sequence.

What does each stage involve?

Title	The investigation should be given a broad name related to the topic you want to investigate, such as weather, rivers or settlement.
Aim	The overall goal/s you wish to achieve should be decided.
Hypothesis	A statement that is to be tested during an investigation should be written. This should be based on a geographical theory.
Planning	Before carrying out fieldwork several aspects of how and where data is collected need to be considered, including: ▶ What data is needed? ▶ Where will the data be collected? ▶ How will the data be collected? ▶ When will the data be collected? ▶ What are the risks involved?
Presentation	The geographical data collected should be presented as tables, annotated photographs, maps or graphs.
Analysis	A written, detailed, examination of a graph, map or table should be carried out to establish trends and anomalies in the data gathered. This should include quoted figures and simple calculations.
Interpretation	Explanations for a pattern and/or trend, in terms of geographical theory and local geographical causes, identified in the analysis. These may be related to theory and local geographical causes.
Conclusions	Decisions drawn about each hypothesis allowing their acceptance or rejection.
Evaluations	A reflective section considering the limitations of the study and possibilities for improvements or further investigations.

▲ **Figure 3** Details of the stages involved in a geographical enquiry

Activities

1. State the meaning of the term hypothesis.
2. List the factors that are considered during the planning stage of a geographical enquiry.
3. Create a rhyme to help you remember the order of a geographical enquiry.
4. Look back through the textbook, find three theories which could be tested.

2 Planning (including aims and hypotheses)

- plan your enquiry by:
  - identifying questions or issues for investigation
  - developing **one** aim
  - developing **a minimum of two** appropriate hypotheses.
- demonstrate understanding of the potential risks involved in fieldwork and how to reduce these risks
- demonstrate understanding of the difference between primary and secondary sources.

How do you identify an issue and develop the aim and hypothesis?

Geographers begin with a topic and then formulate an aim. The aim is then broken down into several hypotheses. A hypothesis is a statement which can be tested using research. Hypotheses are often based on geographical theories. For example, you could develop hypotheses based on the Bradshaw Model shown here (see page 10 for a full explanation of the Bradshaw Model).

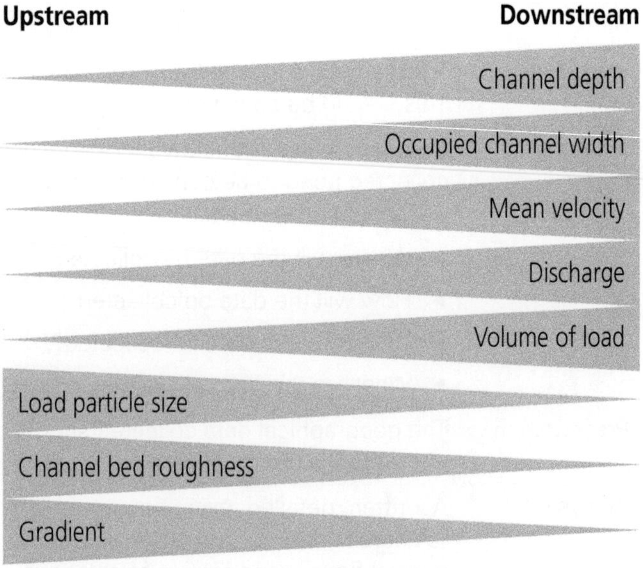

Upstream | **Downstream**

- Channel depth
- Occupied channel width
- Mean velocity
- Discharge
- Volume of load
- Load particle size
- Channel bed roughness
- Gradient

▲ **Figure 4** The Bradshaw Model.

Here is an example outline of a geographical enquiry based on the Bradshaw Model.

Topic: River Environments

Aim: To test the accuracy of the Bradshaw Model on a local river.

Hypothesis 1: River discharge will increase downstream.

Hypothesis 2: Load particle size will decrease downstream.

▲ **Figure 5** Questions you could investigate along a river.

Here is an example outline of a geographical enquiry based on this image of urban land use:

Topic: Settlement

Aim: To test the idea that land use changes along a transect out of a city.

Hypothesis 1: The CBD will have the most commercial land use.

Hypothesis 2: Building heights will be greatest near the CBD.

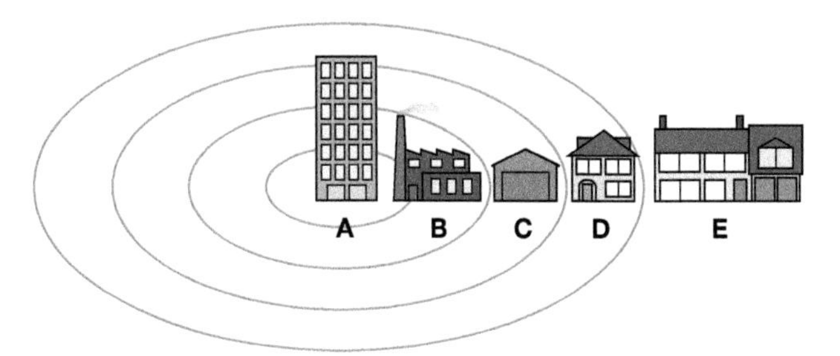

Figure 6 Concentric land use zones in a settlement.

Figure 7 Questions you could investigate in a town or city.

Activities

1 Develop a third hypothesis for the two examples above.

2 Draw out and fill in the table below for each of the following aims.

a To study changes in land use in a town.

b To study the changes in a river channel downstream.

Hypothesis 1	
Info needed	
Hypothesis 2	
Info needed	

Use the examples of hypotheses below to help you.

- Land use in the central business district (CBD) will be mostly commercial.
- The environmental quality in the suburbs is better than that in the CBD.
- Bed load will get smaller downstream.
- Velocity will get slower downstream in the Curley Burn.
- Shoppers have a negative view of the regenerated areas of Downpatrick's CBD compared to their view of the conserved areas.

What are the risks involved in fieldwork?

Before you gather data it is very important that you plan how to stay safe during fieldwork. Your teacher will have completed a risk assessment and will share the most important information with you. However, you are responsible for doing your best to stay safe and follow instructions. Here are some examples of the dangers you and your teacher may need to plan for, depending on the type of study you chose to do.

- deep water
- fast-flowing water
- slippery or uneven surfaces
- steep drops
- traffic
- using public transport
- falling objects
- landslips
- air pollution
- water pollution
- bad weather
- meeting strangers.

Figure 8 Fast-flowing water can become a risk during fieldwork.

Activities

1 Think about the advantages and disadvantages of carrying out fieldwork in the physical environment and in human environments. For example, an advantage of doing an urban study is it may be within walking distance of your school, making it more accessible. Draw out and complete a table to record your ideas, like the one below.

	Physical environment	Human environment
Advantages		
Disadvantages		

2 Write down at least three potential risks to consider before carrying out fieldwork in the following places and state how you could reduce them.
   a in an urban area
   b at the coast
   c along a river

What is the difference between primary and secondary data?

For your geographical enquiry you must use some primary data. Primary data is the information you collect yourself during the fieldwork. You also need to include some secondary data. Secondary data is information that has been produced by someone else.

Collecting primary data

There are several different ways that you can go about collecting primary data in the field. These include:

- Taking photographs or making annotated field sketches.
- Measuring variables using a variety of equipment, e.g. measuring temperature with a thermometer.
- Conducting questionnaires or conducting interviews.

There are two main types of primary data – quantitative and qualitative.

Quantitative data is collected as numbers. It could be the width of a river, the amount of rainfall or number of pedestrians passing a point in a town. The data can then be put into tables or graphs, used in statistical analysis and even compared over time.

Qualitative data often generates descriptive and opinion based results. These can be harder to analyse than quantitative results. Open ended questions in a questionnaire, such as 'What do you think about Belfast?' will generate qualitative data.

Activities

1 Name one source of primary data and one source of secondary data you might use when conducting a river study.
2 Name one source of primary data and one source of secondary data you might use when conducting a settlement study.
3 Suggest why qualitative data is harder to present as a graph and harder to analyse than quantitative data.

By the end of this section you will be able to:

- select data collection methods and equipment that ensure accuracy and reliability
- record measurements and observations accurately using recording sheets
- use at least one secondary source during the investigation.

How do you choose the right data collection method and equipment?

The type of information you need to test your hypotheses will help you decide on how you will gather the data.

Some types of information required:

- height of buildings
- actual land use of buildings in the city centre
- number of pedestrians
- numbers of invertebrates found in rivers
- discharge measurements in cumecs
- speed of river flow in m/s
- origin of shoppers in the city
- scores from shoppers about the regenerated and conserved areas of the CBD
- average axis size of bed load.

Some methods of data collection:

- use a questionnaire to explore shopping habits
- use a flow metre to measure flow at 3 set points across the river channel.
- take a sample of river water and measure the pH
- systematic recording of land use
- pedestrian counts at different distances from the CBD
- photographs of graffiti and litter, or landforms
- use callipers to measure the longest and shortest axes of 10 stones
- measure the width and average depth of the channel, also velocity, then calculate discharge.

Activities

1. Read through the list of examples of types of information required opposite. For each type of information, match it with a suitable piece of equipment and a possible method of data collection.
2. Think about your own title and hypotheses. To test each of your hypothesis write down what equipment you need and how you might use it to collect relevant (useful) information.

▲ **Figure 9** Useful equipment you might consider using in your fieldwork.

How do you record primary data?

Using questionnaires to record primary data

Questionnaires can be used when you want to consult a group of people to find out what their thoughts and opinions are on a particular subject or issue.

These may be physical sheets of paper or you could create a questionnaire online using a survey tool such as SurveyMonkey. Follow the advice below when creating a questionnaire.

- Firstly, work out what information you need to find out. You may be investigating an issue that requires people to have an opinion or perception, or you may be asking them to tell you some specific factual information, such as in an urban study.
- Each question should aim to collect a specific piece of data, which can then contribute towards the final conclusion. Tie each question carefully to a hypothesis.
- Do not have too many questions; even if they are quite brief, people may feel they don't have the time to answer them. Aim for no more than ten questions.
- Start off with a few easy questions to put people at ease. This might include asking them whether they are visitors or residents, for example. However, do not ask personal questions such as age/gender or income.
- Have a mixture of open and closed questions. Open questions enable people to offer any answer or opinion. Closed questions may offer a choice of response, such as 'yes' or 'no', which are useful for later analysis.
- Avoid asking leading questions such as, 'What do you think of the horrible effects of noise pollution?', as this prompts people to answer in a particular way.
- You may want to ask some questions which require people to rate or score something, perhaps on a scale from 1 to 5, or 1 to 10. Be aware that if you have an odd number of scores, there may be a tendency for people to give the middle answer and 'sit on the fence'.
- Think carefully about where you chose to stand when asking questionnaires. Some supermarkets may not like you standing near the entrance to their store. Also, standing outside a particular shop may add a bias to your answers on which supermarkets respondents most often use.

Using mobile devices

Tablets and smartphones are increasingly being used during fieldwork. This could include:

- apps to record sounds or interviews with people
- apps to collect data at particular points, which are then added to mapping tools
- apps to transform or label images (e.g. Skitch)
- video apps to record and then slow down a physical process (e.g. a wave breaking, so that it can be analysed as to whether it is a destructive or constructive wave).

Using a data collection sheet

1 Area visited
Inner City ____________________
Suburb ____________________

2 Complete a tally chart of the building type on the section of transect.

Terraced	Detached house
Flats	Semi-detached
Detached bungalow	Other

3 Make a note of any litter or graffiti noticed within the transect (take a photo for later)

4 Make a note of any traffic calming or parking issues within the transect

5 Complete the following ratings for the transect

boring	1 2 3 4 5	stimulating
ugly	1 2 3 4 5	attractive
threatening	1 2 3 4 5	welcoming
public	1 2 3 4 5	private
drab	1 2 3 4 5	colourful
expensive	1 2 3 4 5	cheap
dirty	1 2 3 4 5	clean
noisy	1 2 3 4 5	quiet
barren	1 2 3 4 5	green
poor	1 2 3 4 5	rich

6 Use the NINIS website to record the ethnic background of residents and the index of deprivation for the transect area.
Ethnic groups by %
Index of deprivation

▲ **Figure 10** Example of a data collection sheet that could be used to compare a 500m stretch of inner city area with that of a suburban area.

Using recording sheets for physical geography fieldwork

When creating a sheet to record your results for physical geography, again you need to focus firstly on your hypotheses and what information is needed to test them. Then create a table or set of tables to allow you to type the data directly into a spreadsheet or record them with a pen or pencil during the fieldwork.

For example, if you are investigating changes downstream in a river and need data on bed load size, the following type of table will help.

▼ **Figure 11** Data recording table.

Pebble	1	2	3	4	5	6	7	8	9	10	11	12	13	14	15	mean
x axis (cm)																
y axis (cm)																

The table below shows a typical way to record results.

Station number	Distance from source (km)	Width of river (cm)	Depth of river (cm)			Velocity (m/s)
			right	middle	left bank	
1	0.5					A B C
2	2.5					A B C
3	4.5					A B C
4	6.5					A B C
5	8.5					A B C
6	10.5					A B C

▲ **Figure 12** Data recording table.

Activities 

1 Explain the difference between primary and secondary data in a geographical investigation.
2 Describe the difference between open and closed questions in a questionnaire.
3 What information might be collected about a river that is not included in Figure 12?

Using secondary data sources

Secondary sources can provide support and offer further insights to help develop your geographical enquiry. These are increasingly digital in nature. The Northern Ireland Neighbourhood Information Service releases population and area statistics on a monthly basis, so you can access very up-to-date data on a local area. You are required to use at least one secondary source of information. This could be a local newspaper report, a weather forecast from the Met Office or even a map.

Using geographic information systems (GIS)

Geographical data will always have a location, which means that the data can be mapped. By mapping the data, patterns often emerge. Maps can be obtained from the trace of a base map or by finding an existing map. You may need to use interactive maps that have been produced for a specific purpose. The Environment Agency, for example, produces flood risk data for specific addresses. This information could be used alongside some practical fieldwork involving questionnaire surveys of householders in areas which have (and haven't) experienced recent flooding.

Tip 

In Unit 3 you might be asked to describe in detail how you collected your data and what equipment you used, so make sure you can spell all the names of the equipment and sources you used.

By the end of this section you will be able to:

- select appropriate ways of processing and presenting your data and explain why you chose these methods.

How do you process data?

Once you have collected your results, quantitative data can be inputted into a spreadsheet or table and simple statistical processing done, such as finding the mean, calculating the range, finding the highest and the lowest values or even simple totalling of scores.

How do you present your findings?

You will be studying and using a range of geographical skills during your GCSE course. Completing the fieldwork enquiry will help you to develop your data presentation and data analysis skills. You can be tested on these skills in all of your exams. In Unit 3 you will need to be prepared to hand draw a graph to represent some of the data you have collected, the second box contains suitable graph types for that part of the examination.

The table below shows some of the ways that you could present your data.

▶ sketch maps of the area ▶ annotated Ordnance Survey maps of the area ▶ choropleth maps ▶ flow-line maps	▶ bar graphs ▶ line graphs ▶ histograms ▶ scatter graphs ▶ pie charts ▶ climate graphs ▶ proportional symbols ▶ pictograms ▶ cross-sections ▶ population pyramids	▶ annotated photographs ▶ sketches and diagrams

Figure 13 Data presentation methods.

Remember that in the examination you will have limited time in which you may have to complete a presentation method for your data, so keep it simple. For example, divided bar graphs can be faster to draw than a pie chart, although both show the proportional breakdown of a set of results.

Markers will not only be giving marks for the accuracy of your graph but also things like having a full and appropriate title, labelling of axes and inclusion of units.

Activities 

1. Look back through this textbook and try to find an example of each of the ways data can be presented as shown in Figure 13. Give a page reference for each one.
2. Look at the data you have collected. Name and explain the best way you can present during the examination.
3. Present the data you have collected.

How do you analyse your findings?

In this section you should be describing patterns or trends in the data using words and figures, as well as noting any odd results. Odd results are known as anomalies.

Look at the graphs or maps you have created from your data. In order to analyse them and work out whether the data supports your hypotheses you'll need to look for patterns and/or trends within the data. When you have identified a pattern, it might be useful to:

- Note if the relationship is positive or negative.
- Calculate at the level of increase or decrease.
- Quote figures to clarify any pattern and/or trend you notice.
- Identify anything which does not fit the trend, i.e. anomalous results.
- Try to make predictions, if appropriate, from the pattern seen in your data.
- If possible, make connections between data sets such as river depth and river velocity.

Figure 14 shows data from pedestrian counts in Coleraine. Here the pattern shows a negative relationship between the number of pedestrians and the distance from the CBD in a town. Point A is an anomalous result as it is much further from the best fit line than all the other results.

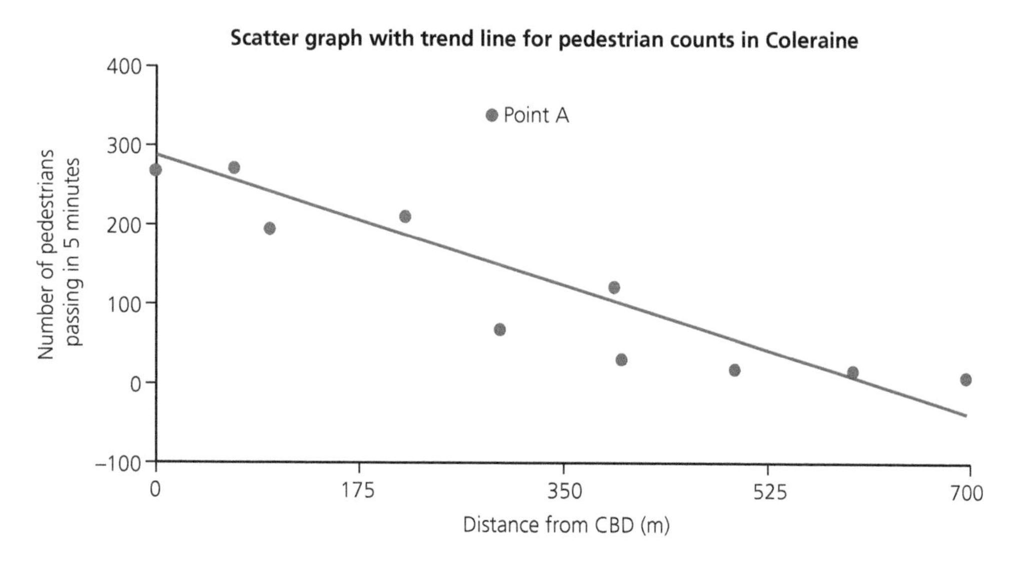

Figure 14 Scatter graph with trend line for pedestrian counts in Coleraine.

By the end of this section you will be able to:

- analyse fieldwork data using your knowledge of relevant theory and/or case studies as appropriate
- interpret fieldwork data using your knowledge of relevant theory and/or case studies as appropriate
- establish links between data sets
- identify anomalies in your data.

Tip

In the exam, when describing patterns or anomalous data you must quote several figures to gain top marks. When interpreting your data, you must refer to theory and local geography within your explanation.

Trend lines on scatter graphs usually suggest a change over time. Here you can see simple examples of positive and negative relationships.

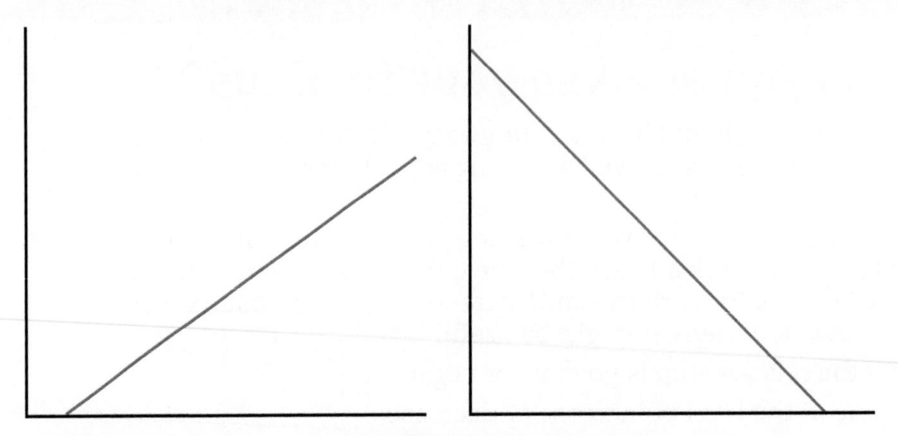

▲ **Figure 15** Graphs showing positive and negative linear relationships.

How do you interpret data?

In this section of a geographical enquiry you must write explanations for the patterns and anomalous results you've found in your data. You'll need to make sure you refer to geographical theory and the local geography.

Patterns in data can usually be linked to geographical theory as part of an explanation.

When faced with an anomalous result you should not ignore it, rather you can often find an explanation based in local geography, such as the location of a bus station which could account for the extra pedestrian count for point A on the scatter graph on page 229.

Activities

1. State what is unusual about result A on the scatter graph on page 229. Why is it an anomalous result?
2. What should you do if you get an anomalous result in your investigation?
3. Analyse (describe) the results of your investigation, focusing on hypothesis 1.
4. Interpret (explain) the results of your investigation, focusing on hypothesis 1.

6 Drawing conclusions and evaluating your fieldwork

By the end of this section you will be able to:

- draw conclusions from evidence
- identify problems with data collection methods
- identify limitations of the data collected
- suggest other data that might be useful
- evaluate your conclusions
- suggest how you could extend the study.

How do you draw a conclusion from your evidence?

When drawing a conclusion you need to make a statement on whether the data supports each hypothesis. Back up your decision with specific evidence from the results you collected. Ensure you make an overall conclusion that refers back to the original aim of the fieldwork.

What are the problems with data collection methods?

Look carefully at how you collected the data, did you have any issues with them when you were conducting the fieldwork?

For example, here are some possible data collection problems during a river study:

- Access issues that meant each point along the river could not be reached.
- Excessive rainfall, which meant the trip had to be postponed.
- Deep or fast flowing water on the outside of a meander.
- Problems reading your map.
- Rain making recording of results harder, possibly reducing accuracy.

Here are some possible data collection problems during an urban study:

- Asking too many questions, or not enough, to answer your hypotheses.
- Being able to get people to stop and answer your questionnaire.
- Being tied to a certain time or day of the week, potentially adding bias to the results.
- Questions on the reliability of secondary sources like newspapers.

What are the limitations of the data?

All data sets will have issues and limitations. It may concern their amount, reliability or clarity.

You need to think about the data you collected, was it enough or do you need more to be able to draw an reliable conclusion? Remember that when there are few results, one anomaly can greatly influence the average.

Are there enough doubts about the results to question the reliability of the data? Maybe you investigated tourism in Belfast and there was a cruise ship in from America and so you found more American tourists than might be expected on a normal day in the city. Would you be sure that if you completed the same work on another day that the results would be repeatable?

Finally are the results clear? If you were measuring the size of a pebble, did everyone measure to the nearest cm or did they give mm? Did everyone use the same key on their questionnaire or did some people make up their own new categories when speaking to the public?

What other data may be useful?

Think about your interpretation section, would it have been easier if you had additional information? Maybe a river study would benefit from having a geological map as well as an Ordnance Survey map.

In urban studies, Address Centred Extract (ACE) maps give very clear detail on the size and location of buildings. They are at a much bigger scale than the 1:50000 used in most Ordnance Survey maps and can show just a few streets with each building outlined. When you are choosing addresses, a system can be useful to order your sampling, such as every 5 metres or every third building. This method of choosing your data sample is called systematic sampling.

How do you evaluate a conclusion?

A conclusion comes towards the end of a geographical enquiry, but it can often lead to further questions being asked, or the conclusion itself might be unreliable. To be truly rigorous, fieldwork should be repeatable and produce a similar data set. If not then the conclusions can be questioned.

For example was there a period of heavy rain during your fieldwork? Did this mean that pedestrian counts might have been reduced? Or was a high pressure system stable over Northern Ireland in the week before your river study, again which might have reduced river discharge below the normal amount for the river you went to?

How can you extend a geographical enquiry?

When looking at any geographical area; river, coast or urban, it is always possible to extend a study further. In river studies two rivers could be compared, or the same river could be compared at different times of the year. In urban studies it could be possible to compare transects out from a town centre to its rural-urban fringe, or try to determine the limit the CBD.

Activities

1 Look at the photos below.

▲ **Figure 16** Photo taken 900 m along the transect.

▲ **Figure 17** Photo taken 1,000 m along the transect.

◀ **Figure 18** Photo taken 1,100 m along the transect.

▲ **Figure 19** Photo taken 2,000 m along the transect.

Activities

1 Now look at the table, which shows information gathered along a transect through the town of Downpatrick, going from south to north through the whole of the settlement.

Distance along transect (m)	Housing	Low order shop	Middle order shop	High order shop	Vacant or derelict
100	0	6	1	0	1
200	0	9	7	0	0
300	0	10	1	2	0
400	0	2	1	3	4
500	0	10	4	1	4
600	0	3	4	0	0
700	0	6	4	1	1
800	0	2	3	1	2
900	0	5	4	1	7
1,000	0	4	1	4	0
1,100	0	6	7	2	0
1,200	1	4	5	3	1
1,300	5	0	1	0	5
1,400	2	0	8	3	0
1,500	2	2	4	0	3
1,600	1	1	0	0	0
1,700	9	1	0	0	0
1,800	2	6	2	1	5
1,900	0	1	2	1	0
2,000	9	0	1	1	0
2,100	4	1	2	2	0
2,200	5	1	1	1	6
2,300	0	0	4	2	3

Low order shops sell items that are low cost and would be bought frequently, such as milk.

Middle order shops sell items which we might compare with other similar items before buying and mostly get used once a month or so, such as clothing.

High order shops sell luxury / expensive goods that might only be bought once a year or less, such as carpet or furniture.

a Does the data support the hypothesis – The CBD in Downpatrick is closer to the south of the settlement of Downpatrick? Draw a conclusion based on evidence.

b Look at the data set, identify any limitations it has.

c Suggest other data which might have been useful.

d Write down a possible extension for this geographical enquiry.

2 Draw a conclusion for each of the hypotheses for which you gathered data – remember to state clearly if the hypothesis has been proven or disproven (or neither) and state results as evidence for your decision.

3 Work in groups to evaluate your fieldwork, the limitations of data, conclusions and state a possible extension. Present your findings to the rest of the class.

Sample examination questions

To help you complete these you must have the A4 sheet with your hypotheses and results table.

1 Copy and complete the chart below by filling in the missing stages needed to complete a geographical enquiry. [3]

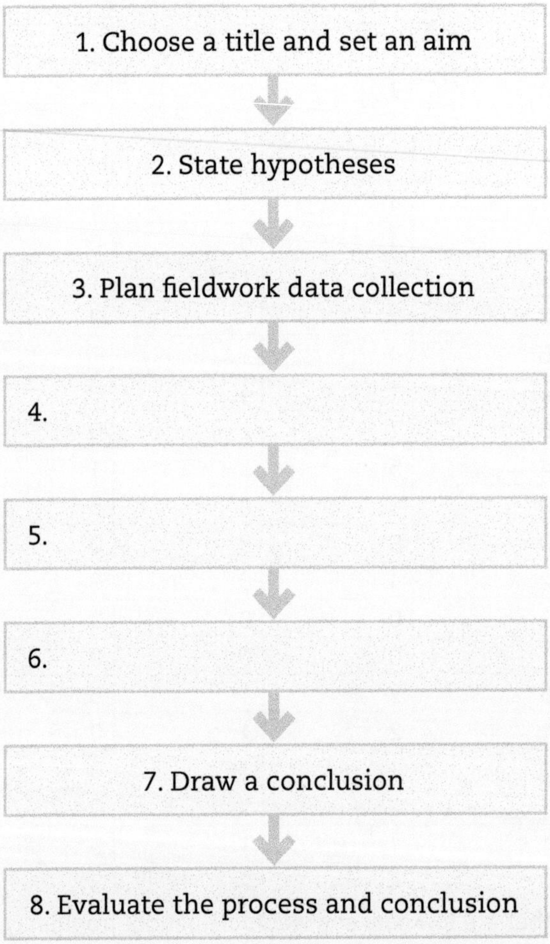

1. Choose a title and set an aim

↓

2. State hypotheses

↓

3. Plan fieldwork data collection

↓

4.

↓

5.

↓

6.

↓

7. Draw a conclusion

↓

8. Evaluate the process and conclusion

2 Explain the suitability of one hypothesis you have chosen to investigate your fieldwork aim. [4]

3 Describe two risks you identified when planning your fieldwork and explain how you tried to minimise them. [6]

4 Name one primary data source you used in your investigation. [1]

5 Describe how you collected the data for one variable shown in your table of results, ensuring you name any equipment you used. [4]

6 Choose one of your hypotheses. Using graph paper present your data relating to this hypothesis. [8]

7 Analyse what your graph shows in relation to your chosen hypothesis. [4]

8 Interpret the results displayed on your graph, ensuring you use geographical theory within your answer. [6]

9 Evaluate two of the methods you used to collect your data and suggest possible improvements. [6]

10 Suggest one way your chosen geographical enquiry could be extended. [3]

Glossary

abrasion / corrosion (1) The grinding of rock fragments carried by a river against the bed and banks of the river. (2) A process of erosion which occurs when a wave hits the coast and throws pebbles against the cliff face. These knock off small parts of the cliff, causing undercutting
afforestation the planting of trees in an area that has not been forested before
aged-dependent the proportion of people aged 65 or over in a population, compared to those of working age (15–64)
air mass a body of air with similar characteristics, e.g. temperature, humidity and air pressure
altitude height above sea level normally given in metres
anemometer an instrument which is used to measure wind speed. It has mounted cups which revolve faster or slower depending on the wind speed
anomaly a result which is unexpected
anticyclone a weather system with high pressure at its centre, generally associated with dry, calm weather
appropriate technology technology which is suited to the level of development in the area where it is used
arches a wave-eroded passage through a small headland. This begins as a cave formed in the headland, which is gradually widened and deepened until it cuts through
atmospheric pressure the weight of a column of air measured in millibars
attrition a process of erosion where transported particles hit against each other making the particles smaller and more rounded
beach nourishment the addition of new material to a beach artificially, through the dumping of large amounts of sand or shingle
barometer an instrument used to measure air pressure
biofuels a fuel that comes from living matter, such as plant material
BRICS Brazil, Russia, India, China, South Africa. Large and growing economies that contribute to global patterns of trade and interdependence
buoy an anchored float which may contain instruments to measure environmental conditions
carbon footprint a measure of the amount of carbon dioxide produced by a person, organisation or country in a given time
central business district (CBD) the part of the settlement which is dominated by shops and offices, often has tall buildings and is usually close to its centre
cirrus a type of cloud, it appears as high, wispy and white and is made from ice crystals
cloud cover the amount of sky covered by cloud, measured in oktas
cloud types clouds are divided into categories (nimbus, cirrus, cirrocumulus and so on) depending on their height, shape and nature
climate the average weather conditions of an area over a long period of time, e.g. 35 years, covering only temperature and precipitation
climate change changes in long-term temperature and precipitation patterns which increasing evidence suggests is caused by human activities
coastal defences these are natural or manufactured management strategies which try to maintain the land-sea boundary or reduce the impact of erosion
coastal landforms large features within a coastal area formed by processes such as erosion and deposition
cold front the zone where cold air comes behind warm air. The cold air undercuts the warm air forcing it to rise, cool and condense
cold sector the larger part of a depression which contains Polar Maritime air
collision margin when two sections of continental crust crash into each other and both are pushed upwards to form a vast mountain range
composite volcano a steep-sided, dome-shaped volcano that erupts a variety of materials such as sticky acidic lava and ash. Occurs at destructive plate margins
conclusion the findings of a study summarised
confluence the point where two rivers meet
conservative margin when plates slide past each other. Crust is neither created nor destroyed
constructive margin when plates are pushed apart, so they move away from one another and new crust is created
constructive wave a wave with a strong swash and weak backwash which contributes deposition to a beach
convention current repetitive movements set up in the mantle due to heating by the core. These currents make the crust move
corrasion an alternative word for abrasion
core the centre of the Earth, found below the mantle. It is extremely hot and may be made of metal
corrosion (1) The grinding of rock fragments carried by a river against the bed and banks of the river. (2) A process of erosion which occurs when a wave hits the coast and throws pebbles against the cliff face. These knock off small parts of the cliff causing undercutting
crude birth rate total number of live births per thousand of the population per year
crude death rate total number of deaths per thousand of the population per year
crust the upper layer of the Earth on which we live. It is solid but is split into sections called plates

cumulonimbus a dense, tall, towering cloud which often has a flat, anvil top. They may produce heavy rain and thunderstorm conditions
cumulus a common cloud type with puffy white tops and often a flat base. They are a low to mid-level cloud
dam a barrier (made of earth, concrete or stone) built across a valley to interrupt river flow and create a purpose built lake (reservoir) which stores water and controls the discharge of the river
data analysis a written, detailed, examination of a graph, map or table to establish trends and anomalies in the data gathered. Figures are quoted and simple calculations carried out
data collection the process of gathering primary information from the field e.g. questionnaire, traffic counts, river velocity recording
data presentation how data is displayed to readers: usually in words, graphs or tables. Geographical data collected should be presented as tables, annotated photographs, maps or graphs
demographer a geographer or scientist who studies populations, particularly of people
demographic transition model a theoretical model based on the experience in MEDCs showing changes in population characteristics over time
depth (river) the distance from the surface of a river to the river bed
deposition the dropping of material on the Earth's surface
depression a weather system with low pressure at its centre, characterised by wet and windy conditions
destination countries countries to which migrants move
destructive margin where one plates crashes into another plate and often crust is destroyed
destructive wave a wave with a strong backwash and weak swash which erodes a coast
development the level of economic growth and wealth of a country. The use of resources, natural and human, to achieve higher standards of living. This can include economic factors, social measures and issues such as freedom
development gap the division between wealthy and poor areas, in particular the disparity between LEDCs and MEDCs
digital thermometer an instrument that measures the temperature of the air, displaying the result as an LCD readout
discharge the amount of water in a river which is passing a certain point in a certain time. It is measured in cumecs (cubic metres per second)
drainage basin an area of land drained by a river and all of its tributaries
earthquake a tremor starting in the crust which causes shaking to be felt on the Earth's surface
economic indicators figures relating to the wealth and economy of a country
ecotourism otherwise known as green tourism. A sustainable form of tourism which involves protecting the environment at the destination
emigration the movement of people away from one country
epicentre the first place on the Earth's surface to feel shockwaves from an earthquake. It is directly above the focus
erosion wearing away of the landscape by the action of ice, water and wind
evapotranspiration the process by which water is transferred from the land to the atmosphere by evaporation from surfaces, e.g. lakes, and by transpiration from plants
fair trade a type of trade where producers in a poor country get a fair living wage for their product and which promotes environmental protection
fault line a weak line in Earth's surface, where crust is moving, causing earthquake activity
flooding a temporary covering by water of land which is normally dry
flood wall a stone/brick/cement wall built alongside a river to protect the nearby areas from flooding
focus the point of origin of an earthquake under the Earth's surface
fold mountain mountain ranges that form mainly by the effects of folding of the earth's crust at destructive plate margins
front the zone where two types of air mass meet
frontal depression a weather system with low pressure at its centre and two contrasting air masses. They are associated with wet and windy weather
function role performed by something; in the case of a city. A function of one part of the settlement might be to provide housing, or employment
gabions metal cages filled with rocks which can form part of a sea defence structure or be placed along rivers to protect banks from erosion; an example of hard engineering
geostationary satellite a satellite that is positioned over one place and moves at the same speed as the Earth. It only provides data for that one place
globalisation the way in which countries from all over the world are becoming linked by trade, ideas and technology
global warming the warming of the atmosphere, i.e. the increase over time in average annual global temperature. This is probably related to human activity through the release of greenhouse gases
gradient the slope over which the river loses height
greenhouse effect a natural process where our atmosphere traps heat. Some of the heat from the sun that is absorbed by the Earth's surface is re-radiated to the atmosphere where it is held by the greenhouse gases – carbon dioxide, methane, nitrogen dioxide, CFCs and water vapour
groundwater flow water which is moving horizontally through the bedrock towards a river or sea

groynes wooden barriers built out into the sea to stop the longshore drift of sand and shingle, and so cause the beach to grow. It is used to build beaches to protect against cliff erosion and provide an important tourist amenity. However, by trapping sediment it deprives another area, down-drift, of new beach material
hard engineering methods strategies to control a natural hazard which does not blend into the environment
headlands areas of land that extends out into the sea, usually higher than the surrounding land; also called a point
hooked spit a coastal landform caused by deposition and the transport process of longshore drift. Changes in the prevailing wind and wave direction can cause a spit to form a curved, hooked end
Human Development Index (HDI) a measure of development which combines measures of wealth, health and education, thus mixing social and economic indicators
hydraulic action (1) A form of erosion caused by the force of moving water. It undercuts riverbanks on the outside of meanders and forces air into cracks in exposed rocks in waterfalls. (2) The process whereby soft rocks are washed away by the sea. Air trapped in cracks by the force of water can widen cracks causing sections of cliff to break away from the cliff face
hypothesis a proposed explanation for something e.g. river load will decrease in size downstream, a statement which is to be tested during an investigation
igneous when referring to rocks, this means rocks formed when molten magma either hardens under the crust or erupts through and hardens on the earth's surface
immigration the inward movement of people to a country from another
infiltration the movement of water into the soil from the earth's surface
inner city areas surrounding a CBD, characterised by mixed land use, with some older terraced residential or newer apartments mixing with industry
inner core the Earth's innermost layer, thought to be solid and made from the metals iron and nickel
interception the process whereby precipitation is prevented from falling onto the ground by plants. It slows run-off and reduces the risk of flash flooding
interpretation making sense of the data. Explanations for trend identified in the analysis. These may be related to theory and local geographical causes
land use zones areas of a settlement which share the same function – such as housing, industrial or commercial
landform a natural, recognisable feature of the Earth's surface
latitude the imaginary lines that surround the Earth ranging from 0° at the Equator to 90° at the poles
LEDCs a less economically developed country, often recognised by its poverty and a low standard of living
levees raised banks along a river that help to reduce the risk of flooding
liquefaction the process of solid soil turning to liquid mud caused by shaking during an earthquake bringing water to the surface
load the sediment carried by a river
longshore drift the process whereby beach material moves along a coastline, caused by waves hitting the coast at an angle
mantle the layer above the Earth's core. It makes up 80% of the Earth's mass. It behaves like liquid rock
mass tourism large scale movements of people to tourist destinations which can result in the building of hotels in vulnerable areas and can have a negative impact on the relationship between local communities and the visitors
meander a river landform. A sweeping curve or bend in the river's course
MEDCs a more economically developed country, often recognised by its wealth and a high standard of living
metamorphic rocks that have been changed as a result of heat and pressure being applied to them over long periods of time
mid-oceanic ridge where two plates made of oceanic crust move apart, the magma of the mantle rises to fill the gap, causing the crust to rise and form a ridge
migration the permanent or semi-permanent movement of people from one place of residence to another. Migration can be classified, for example into forced, e.g. due to war or famine, or voluntary, e.g. looking for better work
millibar the unit used to measure air pressure
mouth the end of a river where it meets the sea, ocean or lake
natural change the difference between the number of deaths and the number of births in a place
natural decrease when the death rate is higher than the birth rate, there is a natural decrease in population. The population will fall, unless there is more immigration than natural decrease
natural increase when the birth rate is higher than the death rate, there is a natural increase in population. The population will rise, unless there is more emigration than natural increase
ocean trench a feature of a destructive plate margin which involves oceanic crust. Where the oceanic crust is forced down into the mantle it sinks below its normal level to create a deep trench in the ocean
oktas a measure of cloud cover, completed as eighths of the sky covered in cloud
outer core a layer of the earth found below the mantle, made from molten metals
overland flow / surface runoff water which is moving over the surface of the land
percolation the movement of water from the soil into the bedrock

plate margin a zone where two plates meet. Plate boundaries may be described as constructive, destructive, conservative or collision
polar satellites these are instruments based in space that orbit the earth as they record information making orbits roughly 14.1 times daily
population pyramid a type of bar graph that shows the structure of a population by sex and age category, and may resemble a pyramid shape
population structure the way in which a population is made up, perhaps by gender or ages
precipitation a form of moisture in the atmosphere, such as rainfall, sleet, snow and fog
prevailing wind the most frequent, or common, wind direction
primary activity an activity which uses the Earth's resources as a way of making money, e.g. farming, fishing, agriculture, mining
primary data data collected by students personally during fieldwork as a result of measurement and observations
primary sources elements in the environment which can be measured using observation and equipment, such as cloud cover or pedestrian numbers
pull factor any attractive/positive aspect or quality of a place which attracts (pulls) migrants to it
push factor any negative aspect or quality of a place which causes people to leave it
quality of life a measure of a person's emotional, social and physical well-being
rainfall radar also known as a Doppler weather radar, this is a type of radar used to locate precipitation, predict its motion, and estimate its type (rain, snow, hail)
rain gauge an instrument which catches and measures precipitation
refugee a person who has been forced to leave their home country and move to another country, often in response to persecution or natural disaster, and has applied for refugee status in the destination country
regeneration the improvement of part of a settlement through rebuilding, restoration or the removal of pollution
renewable energy a sustainable source of electricity production such as wind, solar or biofuels
responsible tourist a tourist who respects the environment and the people of the places which (s)he visits
richter scale a scale between 0 and 9 which measures the strength of an earthquake
risks a judgement of the potential for coming to harm in a given situation
river landforms large scale features found along the course of a river, such as waterfalls and meanders
rock types rocks are divided into categories depending on how they were formed: igneous, sedimentary and metamorphic
rural-urban fringe an area on the outskirts of a city beyond the suburbs where there is a mixture of rural and urban land uses
sandy beach a coastal landform, caused by deposition, which mostly consists of very small mineral particles and has a shallow gradient
satellite image a photograph or remotely sensed image recorded from space
sea walls curved concrete structures placed along a sea front, often in urban areas such as the front of a promenade, designed to reflect back wave energy; an example of hard engineering
secondary data data collected from sources other than the student; may include published material, reports from public bodies and the work of other people
secondary sources source of data collected by others rather than yourself (primary data)
sedimentary rocks that have been produced from layers of sediment, usually at the bottom of the sea
seismograph an instrument designed to measure the energy released by earthquakes
settlement a place where people live and which provides services and places of employment and entertainment
shanty town a characteristic of LEDC cities; an area within them of unplanned poor quality housing which lacks basic services like clean water
shield volcano a wide, low volcano that erupts basic runny lava. Occurs at constructive plate margins
shingle beach a coastal landform, caused by deposition, which mostly consists of medium-sized mineral particles and has a steep gradient
slip-off slope a small feature seen on the outside of some meander bends which is caused by erosion of the river bank
soft engineering methods a strategy to control a natural hazard which does blend into the environment so is often sustainable
solar energy the Sun's energy exploited by solar panels, collectors or cells to heat water or air or to generate electricity
solution the process by which water (in river or sea) reacts chemically with soluble minerals in the rocks and dissolves them
spit a depositional landform formed when a finger of sediment extends from the shore out to sea, often at a river mouth. Caused by the transport process of longshore drift
social indicators pointers, usually of level of development, which are to do with people. Examples include quality of health and education
source the starting point of a river, it may be a lake, glacier or marsh
stacks natural features of an eroded cliff landscape that appear like large free-standing sections of coastline. Stacks are formed by the collapse of a sea arch
stratus a type of cloud that appears as a continuous flat sheet of grey cloud

subduction the sinking of a dense plate into the mantle
subduction zone an area where crust is being forced down into the mantle
suburb a residential area or a mixed use area usually at the edge of a settlement, but within commuting distance of the centre of a city
supervolcano a type of volcano where the potential exists of it erupting at least 1,000km^3 of material, having global consequences
surface runoff / overland flow water which is moving over the surface of the land
suspension the transportation of the smallest load, e.g. fine sand and clay which is held up continually within river or seawater
Sustainable Development Goals (SDGs) 17 global goals with 169 targets adopted by countries around the world in 2015. Spearheaded by the United Nations, these emphasise improving current quality of life while still maintaining resources for the future
sustainable tourism tourism which does not damage the place where it happens, allowing it to continue to be used by future generations
synoptic chart a weather map which shows the weather as symbols over an area
tectonic plate the crust is broken up into seven large sections and various smaller sections, which are floating on the mantle and moving towards, away from and past each other
temperature the hotness or coldness of the air in relation to weather. It is usually measured in degrees Celsius
throughflow water which is moving through the soil
traction the rolling of boulders and pebbles along the river bed
transportation the movement of material across the Earth's surface
tributary a stream which flows into a larger river
tsunami large waves caused by underwater earthquakes
urban planning schemes plans, usually on a large scale, for changing parts of a settlement. Sometimes these change the use of an area from a former industrial area to housing or commercial uses
volcano a cone-shaped mountain built up from hardened ash and lava, from which molten material erupts onto the Earth's surface
warm front the zone where warm air comes behind cold air
warm sector the smaller part of a depression which contains Tropical Maritime air
washlands the river is allowed to flood these areas; it could be farmland or recreational land close to settlements
waste hierarchy the arrangement of waste disposal options in order of sustainability
water cycle the continuous circulation of water between land, sea and air
waterfall a steep fall of river water where its course crosses between different rock types, resulting in different rates of erosion
watershed the boundary between drainage basins, it is often a ridge of high land
wave cut platform a flat area along the base of a cliff produced by the retreat of the cliff as a result of erosive processes
weather the day-to-day condition of the atmosphere. The main elements of weather include rainfall, temperature, wind speed and direction, cloud type and cover and air pressure
weather forecasts a description of the state of the weather in an area with projections on how it might also change in the next few days
width (river) the measurement from one river bank to the other across a river channel
wind the movement of air on a large scale over the Earth's surface
wind direction the geographical direction (compass point) from which a wind blows
wind energy a form of renewable energy, where wind turbines convert the kinetic energy in the wind into mechanical power and eventually into electricity
wind speed the speed at which air is flowing. It can be measured in knots
wind vane an instrument used to measure wind direction using a fixed directional compass and a movable pointer
youth-dependent the proportion of people aged 14 or under in a population, compared to those of working age (15–64)

Index

U

V

W

Y

Z

Acknowledgements

The Publishers would like to thank the following for permission to reproduce copyright material:

Photo credits

p.7 ©Charles E. Rotkin/CORBIS/Getty Images; p.8 ©Hemis/Alamy Stock Photo; p.11 Steph Warren; p.14 ©Charles E. Rotkin/CORBIS/Getty Images; p.17 t ©Chon Kit Leong/Alamy Stock Photo; p.17 b ©David Hiser/Getty Images; p.21 ©Andrew Johns Photography/Alamy Stock Photo; p.22 ©Matt Cardy/Stringer/Getty Images; p.23 ©Darron Mark/Alamy Stock Photo; p.25 ©Atsuko Ellie Teramoto/Alamy Stock Photo; p.29 ©abadonian/iStock/Getty Images Plus; p.33 ©Andrew Stacey; p.37 ©Patryk Kosmider/Fotolia; p.38 t Petula Henderson; p.38 b ©Peter Titmuss/ Alamy Stock Photo; p.39 l ©Nigel Bell/Alamy Stock Photo; p.39 r ©Geogphotos/Alamy Stock Photo; p.40 ©Google Image NASA Image © 2009 TerraMetrics ©2009 Infoterra Ltd & Bluesky Data SIO, NOAA, US Navy, NGA, GEBCO; p.41 ©Adam Burton/Alamy Stock Photo; p.44 ©konstantin stepanenko/Alamy Stock Photo; p.45 l ©Stephen Saks/Lonely Planet Images/Getty Images; p.45 r ©lemonlight features/Alamy; p.46 t ©Andrew Stacey; p.46 b HR Wallingford; p.47 ©Coastal Science & Engineering, Inc; p.48 ©DWImages Northern Ireland/Alamy Stock Photo; p.49 t ©JoeFoxCountyDown/Alamy Stock Photo; p.49 b Courtesy of the National Library of Ireland; p.51 ©Justin Kase z12z/Alamy Stock Photo; p.55 Met Office © Crown Copyright 2017; p.56 tl ©Alexander Tolstykh/Shutterstock; p.56 tc ©Sam Ogden/Science Photo Library; p.56 tr ©Cape Grim B.A.P.S./Simon Fraser/Science Photo Library; p.56 bl ©Paul Seheult/Corbis Documentary/Getty Images; p.56 bcl ©Ashley Cooper/Corbis Documentary/Getty Images; p.56 bcr ©Eyal Nahmias/Alamy; p.56 br ©Philippe Giraud/Sygma/Getty Images; p.57 ©Jeff J Daly/Alamy Stock Photo; p.58 ©Perry van Munster/Alamy Stock Photo; p.60 ©Don Farrall/Photodisc/Getty Images; p.61 ©REDA &CO srl/Alamy Stock Photo; p.62 Met Office © Crown Copyright 2017; p.63 tl ©NASA; p.63 tr ©NASA/Tom Miller; p.63 bl ©Michael Dwyer/Alamy Stock Photo; p.63 br ©Photo-Jope/Shutterstock; p.64 ©Fotolia; p.66 ©barneyboogles/Fotolia; p.69 l ©ian mcilgorm/Alamy Stock Photo; p.69 r ©Stephen Barnes/Alamy Stock Photo; p.71 ©Crown Copyright 2009; p.72 Met Office © Crown Copyright 2017; p.75 ©Handout/NOAA/Getty Images; p.76 ©Odd Andersen/AFP/Getty Images; p.77 ©Simon Roberts/Oxfam; p.81 ©Kyodo News/Getty Images; p.85 ©Aurora Photos/Alamy Stock Photo; p.90 tl ©Dirk Wiersma/Science Photo Library; p.90 tr ©Mark A. Schneider/Science Photo Library; p.90 cl ©E.R.Degginger/Science Photo Library; p.90 cr ©Joyce Photographics/Science Photo Library; p.90 bl ©Aaron Haupt/Science Photo Library; p.90 br ©WILDLIFE GmbH/Alamy Stock Photo; p.93 ©Furchin/iStock; P.94 t ©REUTERS/Alamy Stock Photo; p.94 b ©Kyodo News/Getty Images; p.95 ©2008 Google,Data SIO, NOAA, US Navy, NGA, GEBCO, Image ©2009 TerraMetrics; p.96 ©Romeo Gacad/Staff/Getty Images; p.97 ©Yongrit/EPA/REX/Shutterstock; p.98 ©Danita Delimont/Alamy Stock Photo; p.99 ©NASA; p.102 ©Malcolm Park microimages/Alamy Stock Photo; p.107 ©chapin31/Getty Images; p.120 ©chapin31/Getty Images; p.122 ©Mark Thomas/Alamy Stock Photo; p.124 ©NurPhoto/Getty Images; p.129 ©Fawzan Husain/The India Today Group/Getty Images; p.130 t Stephen Roulston; p.130 b ©Robert Gray/Alamy Stock Photo; p. 131 Stephen Roulston; p.132 Stephen Roulston; p.134b ©Crown Copyright 2017; p.135 ©Crown Copyright 2017; p.137 ©The National Trust Photolibrary/Alamy Stock Photo; p.138 l ©Agencja Fotograficzna Caro/Alamy Stock hoto; p.138 r ©Richard Baker News/Alamy Stock Photo; p.139 ©Jorge Uzon/AFP/Getty Images; p.141 ©Lee Martin/Alamy Stock Photo; p.142 ©Fayez Nureldine/AFP/Getty Images; p.143 ©Daryl Mulvihill/Alamy Stock Photo; p.144 ©scenicireland. com/Christopher Hill Photographic /Alamy Stock Photo; p.145 ©Gareth Duffy/123RF; p.147 t & c Stephen Roulston; p.147 bl ©Tibor Bognar/Alamy Stock Photo; p.147 br ©Tom Irvine/Alamy Stock Photo; p.149 ©imageBROKER/ Alamy Stock Photo; p.153 ©Deshakalyan Chowdhury/AFP/Getty Images; p.160 polyp.org.uk; p.167 Hippo Roller; p.170 ©Owen Franken/Corbis Documentary/Getty Images; p.171 ©David Turnley/Corbis/VCG via Getty Images; p.173 ©Popa Matumula; p.176 ©Sanjit Das/Bloomberg /Getty Images; p.177 t ©Realy Easy Star/Tullio Valente/ Alamy Stock Photo; p.177 b ©Deshakalyan Chowdhury/AFP/Getty Images; p.181 ©REUTERS/Alamy Stock Photo; p.186 ©John Macdougall/AFP/Getty Images; p.188 ©Sergius Ognjannikow/123RF; p.191 ©Niall Carson/PA Archive/ PA Images; p.193 ©Lou Linwei/Alamy Stock Photo; p.194 ©Avalon/Photoshot License/Alamy Stock Photo; p.195 t ©Standard Studio/Shutterstock; p.195 c ©artit/Shutterstock; p.195 b ©liewluck/Shutterstock; p.196 ©frender/iStock/ Getty Images Plus; p.198 ©Jim West/Alamy Stock Photo; p.199 ©REUTERS/Alamy Stock Photo; p.201 ©Fadel Senna/ AFP/Getty Images; p.202 ©COP21/Alamy Stock Photo; p.204 www.cartoonstock.com; p.205 ©Idealink Photography/ Alamy; p.206 ©Rebecca Johnson/Alamy Stock Photo; p.207 ©LHB Photo/Alamy Stock Photo; p.208 ©R.Malasig/Epa/ REX/Shutterstock; p.210 ©Rob Crandall/Alamy; p.212 © Yan Liao/Alamy Stock Photo; p.215 ©MShieldsPhotos/Alamy Stock Photo; p.217 ©idp new zealand collection/Alamy Stock Photo; p.218 Petula Henderson; p.220 ©A ROOM WITH VIEWS/Alamy Stock Photo; p.221 © Ingram Publishing/ThinkStock; p.222 ©idp new zealand collection/Alamy Stock Photo; p.224 tl ©Doug McCutcheon/LGPL/Alamy Stock Photo; p.224 tr ©greenwales/Alamy Stock Photo; p.224 bl ©D.Hurst/Alamy Stock Photo; p.224 br ©Science Photo Library; p.232 Petula Henderson.

Text/artwork acknowledgements

Maps on **pp.18**, **31**, **42**, **52** & **134** reproduced from Ordnance Survey mapping with the permission of the Controller of HMSO. © Crown copyright and/or database right. All rights reserved. Licence number 100047450.

p.43 map from http://www.grida.no/graphicslib/detail/coastal-population-and-altered-land-cover-in-coastal-zones-100-km-of-coastline_7706 © GRID-Arendal 2017; **p.77l** © Philippines Red Cross 2013; **p.77r** adapted from 'Tacloban: a year after typhoon Haiyan' © Guardian News & Media Ltd 2017; **p.108** population data from Country Meters; **p.162** Copper Prices 1989–2012 graph provided by InfoMine; **p.212** Nam Ha visitor numbers and ecotourism revenue 2001–2009 from http://www.tourismmekong.org/wp-content/uploads/2015/06/Nam-Ha-ecotourism-project-Equator-Initiative-Case-Studies.pdf

Every effort has been made to trace all copyright holders, but if any have been inadvertently overlooked, the Publishers will be pleased to make the necessary arrangements at the first opportunity.